PLAIN LANGUAGE

PLAIN LANGUAGE

PRINCIPLES AND PRACTICE

EDITED BY ERWIN R. STEINBERG

Carnegie Mellon University

Wayne State University Press Detroit

95 94 93 92 91 5 4 3 2 1

Library of Congress Cataloging-in-Publication Data

Plain language : principles and practice / [edited by] Erwin R. Steinberg.
p. cm.
Includes bibliographical references and index.
ISBN 0-8143-2020-1 (alk. paper). — ISBN 0-8143-2021-X (pbk. : alk. paper)
1. English language—Reform. 2. English language—Rhetoric. 3. English language—Business English. 4. English language—Technical English. 5. Technical writing. 6. Business writing. I. Steinberg, Erwin Ray.
PE1073.P57 1991
420—dc20 90–38008
CIP

Table of Contents

Introduction: Promoting Plain Language

Erwin R. Steinberg
Carnegie Mellon University, Pittsburgh

The papers in this volume address the problems of promoting "plain language"—language that reflects the interests and needs of the reader and consumer rather than the legal, bureaucratic, or technological interests of the writer or of the organization that the writer represents.

I. PLAIN ENGLISH LAWS

The first section addresses the general concept of "plain language" and various attempts to promote it. The emphasis is on English, although the next-to-last paper concerns French as well, and the last concerns Japanese. Writing to meet the needs of readers is a universal problem, independent of language.

Although seven states in the United States have so-called plain language laws governing consumer contracts and twenty-seven have laws which promote or require plain language in insurance policies, the Federal government's attention to its communication with its citizens and to business's communication with the consumer has been fitful. President Carter issued an executive order designed to simplify Federal regulations, and in the 1970s various government regulatory agencies drafted regulations requiring "readable" patient package inserts (for various medicines and drugs) and warranties; but Federal attention to such matters has languished in the 1980s.

Although, as Bowen, Duffy, and I report, fifteen states have plain language legislation pending, state attention has languished, too. In Pennsylvania, for example, plain language legislation was drafted in 1981 and introduced in the state legislature in 1983, but the bill has met continued opposition. Interest in plain language, however, is being kept alive by the kinds of groups represented by the authors who have contributed to this book: companies convinced that good communication is good business, plain-language research and consulting organizations, and various departments in universities. It will probably take a resurgence in the consumer movement and more action at the Federal level to persuade state legislatures to pay serious attention again to plain language legislation.

In the paper that opens the first section, Carl Felsenfeld, who was primarily responsible for Citibank's leadership in the use of plain language in the 1970s and is now a professor at Fordham Law School, reports on the effects of the Sullivan law, the first "Plain English" law (New York State, 1978), on the New York marketplace. Instead of the difficulties widely predicted by lawyers when the law was passed, the Sullivan law has resulted in massive simplification of New York consumer contracts and has resulted in almost no litigation. In the next paper, Betsy A. Bowen, Thomas M. Duffy, and I describe briefly the interest of the Federal government in plain language and the variety of plain language laws passed by seven states. We then outline a method for evaluating the existing plain language laws to determine which of them is most cost-effective.

The last three papers in section 1 report on plain language activity outside of the United States. Robert Eagleson, of the University of Sydney and Special Advisor on Plain English to the Australian Government, describes the development of the plain language movements in Australia and the United Kingdom. In both countries, insurance companies led the way, followed soon by consumer groups and the government. In Canada, Gail Dykstra reports, plain language activity was begun independently by an insurance company and a bank, and another group is now taking a leadership role as well—the Canadian Law Information Council, which has assumed responsibility as a catalyst in promoting plain language. (The paper by Irene Etzkorn in section 4 demonstrates that provincial government in Canada is also employing concepts of plain language in communicating with taxpayers.) Hiraku Amemiya, Document Design Manager for High Technology Communications in Tokyo, explains that the plain language movement came somewhat later to Japan than to the various English-speaking countries and explores some of the challenges facing technical writers in his country. As shown in the Dykstra, Etzkorn, and Amemiya papers, although plain language problems may differ from one country and from one language community to another, the principles involved in solving them are the same: training writers to be sensitive to the needs of their readers, and giving writers the skills to satisfy those needs.

II. THE LANGUAGE OF PLAIN ENGLISH

From his experience working as a consultant in business and government, Joseph William, of the University of Chicago, believes that at the level of sentence structure we know 90% of everything people who write need to know in order to write plain English. He sees the major problem as learning how to teach the information, getting people to use it, and persuading the highest levels of management that actively promoting clear communication should be their highest priority. In the process, he offers some guidelines for effective writing. Thomas N. Huckin and Elizabeth H. Curtin, of Carnegie Mellon University, and Debra Graham, of Software Architecture and Engineering, argue that in an attempt to provide useful guidelines for writers, proponents of plain English sometimes offer maxims that oversimplify and do not reflect the actual practices of good writers. They demonstrate by considering the maxim "Avoid whiz-deletions" (reduced relative clauses).

III. AIDS TO WRITING, REVIEW, AND REVISION

It is not enough, of course, just to shake a finger at writers and chide them for not writing "in plain English." What sort of advice and support is available to them?

In the first paper in this section, Janice Redish and Susan Rosen, of American Institutes for Research, pick up the problem raised by Huckin and company and examine the benefits—and dangers—of establishing plain language guidelines. In the process, they disagree with the Huckin-Curtin-Graham paper and with some others who have written on guidelines. Clearly then, how plain language guidelines can best be used is still a matter of controversy.

The next three papers discuss how the growing use of computers for word processing has encouraged the development of computer programs that aid in composing and editing. Lance Miller, of IBM, describes that company's EPISTLE, and Lawrence Frase and Nina Macdonald, of AT&T's Bell Laboratories, describe their company's Writer's Workbench.

The final two papers, the first by Karen A. Schriver and the second by Joyce Hannah Swaney, Carol J. Janik, Sandra J. Bond, and John R. Hayes, show how protocol-aided revision is used in the Communications Design Center at Carnegie Mellon University to evaluate and revise draft documents and to clarify text for particular audiences. The authors' work demonstrates that proponents of plain language must do more than promote the use of simple language; however important it is to teach writers to pay attention to style and the surface features of language, it is also important to test readers' understanding before approving a document for use.

IV. CASE STUDIES

The final section describes specific plain language projects and programs in government, commerce, and industry. In separate papers, Barry Jereb, of Allen-Bradley, and I discuss how the Industrial Computer Group of Allen-Bradley and the Communications Design Center of Carnegie Mellon University mounted a program to improve the language and design of Allen-Bradley documentation. We each conclude that, however much experts disagree on the value of guidelines, companies feel that such guidelines are useful. Consequently, they seek help in establishing guidelines for specific writing situations and ask their writers to follow them.

In her paper, Irene Etzkorn of Siegel & Gale, a design and language simplification firm, discusses problems with government forms and documents and then presents several case studies which demonstrate ways of attacking and solving those problems. Ronald G. Amodeo, Michael J. Floreak, and Lorraine K. Wilkin, of Fitch RichardsonSmith, another design and language simplification firm, demonstrate that plain language can be not only an effective sales tool for the manufacturer, but also a service to the customer purchasing a product to serve his or her particular needs.

The last paper shows how good design can be used as a guide in writing to provide effective communication. Reed Agnew, of Agnew Moyer Smith, a design firm, discusses the development of a report on an integrated information system—often referred to as "the office of the future." Instead of starting with the text and then adding visuals, Agnew first developed a visual model and then wrote the report.

In the process of developing their various topics, the authors discuss a variety of forms and documents in business, industry, and government: a guarantee, a report, an employees' handbook, an insurance policy, computer manuals, a public recall notice, consumer information printed on and in packages, food stamp forms, census forms, and tax forms.

A recurring theme that runs through many papers and a matter that should lead to continued debate and research is the use of guidelines: are they useful? for whom? can they be misleading? how should they be written? As more and more of us have learned to use word processors to write, that debate has already been broadened to include the question of whether guidelines should be supplied "on line" (i.e., on the computer screen) or "off line" (i.e., in "hard copy"—book—form). None of the papers in this volume raises that question, but there is already a voluminous—and growing—body of research seeking to answer it.

Many of the papers in this book were given at a seminar in Plain English which was held at Carnegie Mellon University in March 1984 and cosponsored by that university and The Plain English Forum. Several papers were added to that group for a special issue on "Promoting Plain English" in *Visible Language* (vol. XX, Spring 1986). For this book, several additional papers were solicited to provide further discussion of the principles, problems, and advances in promoting and using plain language. In the special issue of *Visible Language,* Merald E. Wrolstad, the late editor of that journal, allowed each author to use the style of footnoting and bibliography with which he or she was most comfortable. I have held to that policy for this book.

1

PLAIN LANGUAGE AND PLAIN LANGUAGE LAWS

The Plain English Experience in New York

Carl Felsenfeld
Fordam Law School

New York State enacted its Plain English Law, the first in the country, on August 5, 1977. Derived largely from an innovative promissory note written by Citibank, the law, because of its vague standards of clarity, raised questions in the New York legal community that any contract might be deemed "unclear" or "incoherent." Very few cases have arisen, however, and contracts have not lost their validity. The dominant reason for this is probably the low penalty structure in the statute. Lawyers are not stimulated by that structure to bring class actions whose main goal is the payment of legal fees by the losing defendant. At the same time, contracts in the state have been largely rewritten and are vastly more understandable to the consumer than they had been before the law. Other plain English laws of general application have been enacted in six other states, and the Commissioners on Uniform State Laws have been asked to draft such a law for national enactment.

On August 5, 1977, New York State became the first state in the United States to enact a general-purpose plain English law.[1] The derivation of the law requires some explanation.

The basic source of the statute was a plain English promissory note introduced by Citibank on January 1, 1975. Prior to this note, such documents issued by banks were almost entirely drafted in traditional legalese. That is, bank counsel would be asked to write a promissory note. They would respond with a "carefully" drafted document. The standard method was to take the promissory note last used—the "precedent" as it was called in the law office—and add to it verbal solutions to the problems that had been encountered, or even divined, since its most recent publication. Thus, promissory notes grew increasingly longer and more complex as used. The classic lawyer is much better at addition than subtraction.

It made no particular difference that the note was to be used in consumer rather than commercial transactions. It also made little difference that the note would be used to evidence a transaction of, say, $500 rather than, $5 million. A promissory note was a promissory note, and that was that.

In the early 1970s, it occurred to a small group at Citibank that it might well be useful to write a document for the small, relatively uncomplicated and typical consumer transaction that would, first, contain only what it had to contain, and second, say it so that the consumer-borrower could understand. This approach, essentially taken for granted today, was greeted with indignation by both bank lawyers and bank executives. The former group expressed concern that massive consumer outstandings in favor of the bank would be put at risk through the use of new and untested language. The latter group saw no particular benefit either to the bank or to the consumer-customers. Some of the arguments most regularly heard included: consumers choose their bank because of its convenience, not because of the way its forms are worded; consumers don't read their documents; if it ain't broke, don't fix it.

Ultimately, the lawyers gave their grudging consent and the bank unified its energies behind the new form. The national, even international, acclaim that greeted the form surprised everyone involved with its preparation. Television and the national press saw it as a major event. Regardless of its benefit either to the bank or to the consumer, points that are still debated, it was viewed as a major benefit by consumer spokesmen, from the Chairman of the New York City Department of Consumer Affairs to Senator William Proxmire, Chairman of the Senate Banking Committee.[2] The small renegade group at Citibank had clearly accomplished what it had intended to do.

Affairs, however, soon got out of hand. A cub assemblyman in the New York legislature introduced a bill to the effect that all promissory notes used with consumers in the state had to be in the Citibank form unless the Banking Department authorized otherwise. The bill included a full text of the Citibank note, which, due to the enthusiasm of Yonkers assemblyman Peter Sullivan, continued to contain the name of Citibank's insurance carrier. Though flattered, Citibank nonetheless announced that it saw only mischief if everyone, including even Citibank itself, were forced to use that single form, and the bill was killed.

Undaunted, Peter Sullivan soon introduced a new bill, very similar to the ultimate statute, which contained two basic requirements for consumer agreements:

1. They had to be written in nontechnical language and in a clear and coherent manner using words with common and every day meanings; and
2. They had to be appropriately divided and captioned.

Duncan McDonald, a Citibank officer largely responsible for the original note is reported to have been the draftsman of this bill. Largely due to lack of attention, the Sullivan bill moved quickly and easily through both houses of the New York legislature. (In the process, the "nontechnical language" prohibition was deleted through fear that one could not, for example, buy an "antenna.") Opposition did not have a chance to galvanize until the bill appeared on the governor's desk for his signature. At that point, opposition from the business and legal community was almost unanimous.[3]

Dubbed by lawyers a "second Sullivan Law," the New York bar maintained that, should the governor sign it, the law would turn all contracts into potential hidden weapons. Lawyers pointed to the already crowded judicial calendars and conjectured long lines of consumers and their lawyers ready to attack the language of their agreements.

Citibank itself opposed signature by the governor, pointing to the absence of regulation that led to the inventive form in the first place. The company hoped to continue its innovative work in this area and feared that any judicial decision distinguishing between clear language and traditional legalese would of necessity impede future creativity.

The governor, however, pressed by his consumer advisers, acceded to the will of the legislature. Using the occasion for some plain language of his own, the bill became law accompanied by the following governor's memorandum:

> What does the bill do?
>
> The bill requires all consumer contracts to be:
>
> a. Written in plain English.
>
> b. Appropriately divided.
>
> c. Properly captioned.
>
> What is a consumer contract?
>
> A consumer contract is an agreement about money, goods or services valued at less than $50,000.
>
> When does the bill take effect?
>
> The bill takes effect on June 1, 1978.
>
> I approve the bill.
>
> Hugh L. Carey

The statute was subsequently amended to extend the effective date to November 1, 1978. Banks, retailers, small loan companies, landlords, form printers, and the many other businesses affected by the law needed time to revise their forms. The Sullivan concept, however, became the law of the state of New York on that date.

As a result, throughout the state, forms used for consumer transactions were changed. Traditional verbose legalistic language was either eliminated or reduced to a form that became considerably more understandable to the consumers who signed the forms. There was next to no litigation.

A few cases did arise. Two tested the relationship of the Plain English Law with the New York State Rent Control Law.[4] The former required new language for new contracts; the latter required that leases be renewed in the same form as before. The technical conflict was worked out. In another situation, the attorney general of the state of New York started an action against the Lincoln Savings Bank for a verbose and incomprehensible contract relating to its safe deposit boxes. The chairman of the bank asserted that he should have been telephoned rather than sued. He changed the form and the action was dropped. This placid situation continues to exist today.

Why, then, when catastrophe had been predicted by the united New York bar, did the sky not fall? To understand this, we must return to the New York statute.

First, the statute is clear that obligations even in traditional verbose legalistic form are still obligations.[5] One is not relieved of his duty to repay a loan or pay rent because the contract is not understandable. Such a violation gives rise only to an action for the prescribed penalties under the plain English law. Those seeking a lightening of their legal burdens get no benefit from Peter Sullivan and, therefore, bring no lawsuit.

Second, and by far most important, is the penalty structure of the statute. In discussions concerning his bill during its legislative phase, Sullivan described his objectives as being essentially moderate. He said that he was looking for improvement in consumer contracts; he was not seeking the ultimate in understandability. With this in mind, he made his penalty structure relatively moderate. A single violation would cost the perpetrator $50. Class actions could not rise above $10,000. There would be no award of attorneys' fees to successful consumer plaintiffs.

While the class action $10,000 ceiling may appear, at first blush, to be a relatively hefty amount, it is not nearly enough to attract the attention of the New York strike suit bar. Compared to the limitless penalties plus counsel fees that were available under the original Truth-in-Lending Act, it is barely worth a glance. The massive judgments for niggling violations that were becoming available under the Truth-in-Lending Act were, indeed, a warning for Peter Sullivan in drafting his statute. (The Truth-in-Lending Act was amended in 1979 to put ceilings on the penalties and floors under the violations and has, since that date, resulted in a much more reasonable pattern of enforcement.)

The point is that there was little dollar incentive in the penalty structure of the New York plain language law. Thus, despite the dire warnings of the New York bar, actions were not brought. On the other hand, the businesses of the state, lenders, sellers, form printers, and others, do not essentially concern themselves with the degree of penalty when a new statute is enacted. The level of penalty may indeed be a factor when subtle risks exist and decisions are made as to the degree of compliance. But a department store or a bank, for example, doing business in the state of New York will not simply ignore a statute regardless of its penalty level. It will comply; and most did comply. Barring such aberrations as the Lincoln Bank tale, forms were changed throughout the state to a higher level of consumer understanding.

The penalty level did create, however, some problems of its own. The Committee on Consumer Affairs of the Bar Association of the City of New York expressed its concern that the level of compliance might not be as high as it would have been in a more punitive legislative atmosphere.[6] It agreed, however, to live with the Sullivan statute to see how events would transpire. It has never taken subsequent action.

There may indeed have been an element of validity in the committee's original position. The Landlord's Association of New York, for example, redrafted its forms to comply with the plain language law. These forms are widely used by the landlords of New York State and, without question, are considerably simpler than prior versions. The new forms have been criticized, however, for not being as consumer-oriented as they might be. Nevertheless, they are a significant improvement over what they had been. This was Peter Sullivan's objective. Based on that standard, the Sullivan bill must be considered a complete success.

Perhaps at some point we will understand language better than we do today. Perhaps we will understand why some messages are clear to consumers and why some are not. On that day, we might wish to strengthen Sullivan's penalties and perhaps impose our standards upon such organizations as the Landlord's Association. Until then, however, the Sullivan approach must be perceived as a reasonable method of meeting a difficult problem.

We cannot precisely define words such as "clear" or "coherent," words that are the key of the Sullivan law. The modest penalty structure does represent one method of dealing with that enigma. Laws built on cloudy language are far from unknown to the American legal system. The government may not, for example, take property without "due process of law."

Our description of the New York plain language law might well end here. Two developments are, however, worth discussing. First, in New York State, the State Law Revision Commission appreciated the theoretical difficulty inherent in a statute whose key words are not subject to generally available definitions. Consequently, in 1978 it recommended a general overhaul of the Sullivan statute with a series of helpful definitions. Happily, the legislature rejected the effort and we continue to live with Peter Sullivan's brilliant solution to the problem.

Second, we have the experience of other states. Plain language requirements appear in the laws of many states. Some twenty-six states have statutes mandating that insurance policies be written in plain language.[7] Unlike the Sullivan approach, those statutes typically adopt the so-called Flesch test, named after its developer, Rudolph Flesch, a Viennese-American with strong language credentials. The Flesch test rates text readability on a scale from 0 to 100, with 100 being the easiest to understand and 0 the most difficult. The test is based on the length of words and the length of sentences and is particularly useful for longer documents where those two qualities become more meaningful. For short consumer contracts, however, the test loses validity. For example, the sentence "Kick the cat" scores exactly the same as "Cat kick the."

Plain language laws of general application have been adopted in only six states in addition to New York.[8] Bills introduced in well over half of the states since adoption of the Sullivan law in New York, however, have generally been defeated through the opposition of the entrenched bar and business interests. Occasionally, one still learns of a bill introduced here or there but interest in plain language legislation is simply no longer the consumer fashion. Unfortunately, those states that have enacted plain language laws have experienced more concern with the theory that the Sullivan law approach cannot work than with the fact that it does. These new laws are, therefore, filled with legal contrivances designed to solve the problem of compliance. Technical tests, Flesch scores, and even resort to state administrators is the general rule of these statutes. They have, however, worked about as well in practice as has the Sullivan statute in New York.

The move toward plain language legislation is not, however, at an end. A national, uniform law of plain language has been proposed, but it is not, at this time, being taken seriously by the National Conference of Commissioners on Uniform State Laws, whose approval of such a device is fundamental to its general acceptance. In the future, Peter Sullivan and his experience in New York could well serve as the template for successful plain language legislation.

1. The law is codified at N.Y. Gen. Obls. Law, Sec. 5–702.

2. See 1975 *Congressional Record,* p. 2110.

3. See "Plain-English Law in Need of an Interpreter," *N.Y. Times,* Oct. 29, 1979.

4. See *Newport Apartment Co. v. Collins,* N.Y.L.J., May 16, 1979, at 13 (Civ. Ct. N.Y.) and *Frances Apts. v. McKittrick,* N.Y.L.J., June 6, 1979, at 11 (Civ. Ct. N.Y.).

5. See Section b of the statute.

6. See Committee Report dated March 1977.

7. See Felsenfield and Siegel, "Simplified Consumer Credit Forms," (1986 Cum. Supp.) Warren, Gorham & Lamont, p. S–xxxii.

8. In addition to New York: Connecticut, Hawaii, Maine, Minnesota, New Jersey and West Virginia.

Analyzing the Various Approaches of Plain Language Laws

Betsy A. Bowen, Thomas M. Duffy, and Erwin R. Steinberg
Carnegie-Mellon University, Pittsburgh

Plain language laws, designed to ensure that consumers can understand and use the personal business contracts they sign, have recently become an important new kind of regulatory legislation. Seven states have passed laws to control the comprehensibility of consumer contracts; fifteen states have similar legislation pending. These plain language laws range from a simple statement of a general principle to extensive and explicit requirements. Although plain language laws have been in effect for over seven years, no comprehensive evaluation of their effectiveness has yet been done. We propose an evaluation of the plain language laws to determine the characteristics of language and design associated with greatest ease of use, and the design strategies of companies and designers which produce the most usable and comprehensible contracts. This analysis would enable us to identify the best model for plain language legislation. Such an evaluation of plain language laws would require two phases: an analysis of contracts and consumers' ability to use them, and an analysis of the design process which led to the contracts that are easiest to use.

State and Federal governments enact substantial amounts of regulatory legislation to aid consumers or to correct apparent injustices for consumers. Much of this legislation — automobile safety regulations, for example, and environmental protection legislation — has produced real benefits. Unfortunately, however, there is a drawback to all regulatory legislation: it is expensive. The legislature spends time and effort to draft the laws. Industry similarly must spend time and effort to comply. This expense means that policy makers need to weigh carefully the costs and the benefits of legislation before they adopt it. When it is in place, policy makers need to review it to ensure that it is operating effectively, that it helps the consumer and does not unnecessarily penalize industry.

Recently plain language laws have become an important new kind of regulatory legislation. These laws were designed to put consumers on a more equal footing with industry when they sign personal business contracts. Traditionally, contracts put consumers at a disadvantage because such contracts were often filled with lengthy, complex sentences and unfamiliar terms which the consumers could not understand. Moreover, some of these contracts were written in small print with no index or headings to help the reader find information. Consumers, therefore, sought legal remedies through their legislatures.

Reprinted with permission of the publisher, *Visible Language,* Rhode Island School of Design, Providence, RI 02903.

Simplifying consumer contracts benefits the consumer by enabling him or her to identify the rights, obligations, and remedies that the contract provides. Furthermore, simplification also benefits industry by increasing competition in the marketplace and increasing consumer trust and goodwill toward business and industry (PCCC 1983).

The Federal Impetus

Plain language legislation grew out of the consumer movement which began in the 1960s. As consumers grew increasingly skeptical of government and big business, they began to demand contracts that they could read and understand (Clive and Russo 1981: 208). The so-called plain language movement grew in the 1970s and gained national attention in March 1978 when President Carter issued Executive Order 12044, designed "to make Federal regulations clearer, less burdensome, and more cost effective" (Carter a 1978: 561). It required that any regulation issued by Executive Agencies must be "as simple and clear as possible, written in plain English, and understandable to those who must comply with it" (Carter b 1978: 558). In his "Statement on Executive Order 12044" President Carter compared the original and a revised version of a regulation governing CB radio owners to provide an example of the kind of change the order was designed to bring about. The original read: ". . . applications, amendments thereto, and related statements shall be personally signed by the applicant, if the applicant is an individual." The revised version simply said: "If you are an individual, you must sign your own application personally" (Carter a 1978, 562). The two provisions mean the same thing, but the second is unquestionably clearer to most people.

There were two very visible responses to that order. First, the Internal Revenue Service extensively revised the income tax form 1040A to make it easier to use and understand. Second, the National Institute of Education contracted with the American Institutes for Research, the Document Design Center of Carnegie-Mellon University, and Siegel and Gale to provide assistance to government and industry in revising and developing plain language documents; to develop a university curriculum which provides instruction in the principles of document design for the user; and to conduct research in the area of plain language document preparation. Under this contract, a document design workshop for industry and government employees was developed, extensive assistance and consultation in the design of documents was provided to industry and government, and an undergraduate course and a graduate program were developed.[1]

There does not appear to have been a widespread response, however, to the Carter executive order, probably for a variety of reasons. First, neither the order itself nor the statement that the President issued with it set out any criteria for judging whether a document was written in "plain English." Statements in the order about the language to be used in government documents were very general. For example, Carter wrote that Executive Order 12044 "will direct that regulations be written in plain English. Government regulations are usually written by experts for experts. Your clear mandate will be to translate regulations into language a small businessman—who must be his own expert—can understand" (Carter a 1979: 561).

Second, as a former regulation writer explained, the people who write government regulations do not recognize that they speak and write a special language that is foreign to the average citizen: "Those of us who live and work in Washington have our own patois, our own sort of bureaucratic cant. It is easy to use, and we use it every day. Some of us even love it. This is fine, I guess, as long as the only people with whom we attempt to communicate speak governmentese as fluently as do we. Bureaucrats, however, tend quickly to forget that 99% of the people don't speak governmentese..." (Jones 1980).

Third, the executive order covered only the regulations written by Executive Agencies — the Environmental Protection Agency, for example, and the various Federal departments such as Education and Transportation. It had no effect on regulations written by independent agencies like the Federal Trade Commission, the Federal Communication Commission, or the Occupational Safety and Health Administration. That meant that the order governed only a small portion of the regulations "a small businessman" — or any other citizen — has to deal with.

Fourth, President Carter's executive order did not have much time to have any effect on the Federal bureaucracy because three years after it was issued it was revoked by Section 10 of President Reagan's Executive Order 12291, issued in February 1981 (Reagan 1981: 130). Federal support for research into applications of plain English dwindled, as did interest in many government agencies in the use of plain English.

Finally, of course, President Carter's executive order had no impact on communication in business and industry — on such documents, for example, as warranties that came with mechanical items, patient package inserts that came with drugs, instruction manuals that came with unassembled toys or computers, or loan contracts from banks and lending companies. President Carter's interest in plain language and his remarks about it (some of them reported in the newspapers) did evoke some interest in the business community. A number of businesses, most notably Citibank, recognized the need for plain language contracts and the possible advertising advantages such consumer-oriented actions could have.

In the Reagan Administration, Secretary of Commerce Malcolm Baldridge has been active in trying to persuade businessmen of the cost effectiveness of plain English. He arranged for a meeting at the Department of Commerce to consider the usefulness of plain English in business and ways in which it could be promoted. He explained: "As far as the business community is concerned, the best way to discourage bad habits is to emphasize the profits that can result from simple language. The plain English programs of your companies prove this. Your well-written contracts, warranties, forms, insurance policies, and information brochures are helping you to serve customers better and improve productivity" (Baldridge 1983: 1).

Out of that meeting came the Plain English Forum, an organization of business people, academics, and other research people interested in promoting plain English in business. The U. S. Department of Commerce has also published, through the U. S. Government Printing Office, *How Plain English Works for Business,* twelve case studies which "describe how some business organizations have scored successes by simplifying consumer documents [like] warranties, credit contracts, insurance policies, and product information booklets" (Office of Consumer Affairs 1984).

State Laws

The greatest influence of the plain English movement, however, has been on the development of state legislation to regulate the clarity of consumer contracts. Twenty-eight states have passed legislation to control the readability and, therefore, the usability of life, property and casualty, and health insurance contracts. This legislation has focused primarily on controlling the word length and sentence length in the contracts, key variables in all readability formulas. Seven states — New York, Minnesota, Hawaii, West Virginia, New Jersey, Maine, and Connecticut — have passed legislation aimed at controlling the comprehensibility of consumer contracts. Another fifteen states have comprehensive legislation pending.

The plain language legislation in all these states has a common objective: to ensure that citizens in the state are able to understand the rights, obligations, and restrictions in any consumer contract they enter into. The only way, of course, to ensure that a contract is understandable is to test consumers' ability to understand and use it. States have been reluctant, however, to require such user-testing, since testing all revised documents this way would be both expensive and time-consuming. Instead, the legislators have searched for ways of specifying the characteristics of a document which will reflect ease of use and understanding.

There have been two primary issues in the attempt to legislate plain language criteria. First and most obviously, documents which meet the criteria must be easy to understand and use. That is, the criteria specified for evaluating documents must be a good substitute for testing of the document by actual users. Second, to encourage compliance by industry, the criteria must not impose an undue or unnecessary burden on developers of documents. The response to these requirements, and hence the nature of the plain language laws, has been very different in the seven states. Furthermore, the range and diversity of requirements will probably increase when others of the fifteen states with pending plain language legislation pass such legislation.

The 1978 New York law, the first plain language law, represents one extreme in plain language law drafting. It simply states that any contract it covers must be "written in a clear and coherent manner using words with common and everyday meanings, [and] appropriately divided and captioned by its various sections" (New York General Obligations 1978).

Thus "plain language" is defined only in general terms referring to vocabulary, writing style, and headings. The law, which provides no objective standards, thus reflects a view that plain language standards cannot be dictated. Producing a plain language document is a matter of orientation, of placing oneself in the position of the intended reader. The particular characteristics of the plain language document will depend on the particular document and the particular designer — many "plain language" designs are possible. Many who hold this point of view believe that legislating objective criteria may result in compliance with the letter of the law, (i.e., the criteria), but may not result in compliance with the spirit of the law (i.e., usable documents).

The Connecticut law represents the other extreme. This plain language law, the most directive of the seven such laws, provides specific requirements for the length of words and sentences, the type size, the use of headings, acceptable terminology, and organization. It requires that the average number of words per paragraph be less than 75, the average number of words per sentence less than 22, and the average number of syllables per word less than 1.55. The law also requires that each section be captioned in at least ten-point type, that the body of the text be written in at least eight-point type, and that there be margins of at least one-half inch on all borders of the page. The law reflects the view that "plain language" can be defined objectively and that any document will be comprehensible if it meets specific requirements and the designer makes a good-faith effort to write a clearer contract. This legislation thus presumes that compliance with the intent of the law is easier to achieve if it specifies objective, measurable criteria.

Of the five other states that have passed plain language legislation, four — Minnesota, Maine, Hawaii, and West Virginia — have adopted the wording of the New York law verbatim. Minnesota and Maine have supplemented their laws with an optional review of revised documents by the state attorney general. Thus, in these two states, in contrast to the New York law, an official assessment of compliance can be made without litigation. New Jersey uses the general principle contained in New York's law, but adds a list of some of the features the attorney general may consider in reviewing a document.

The need for evaluation

Although there have been plain language laws for over seven years and although laws are being considered in an increasing number of states, there has never been a comprehensive evaluation of the effectiveness of such legislation. Since the goal of plain language legislation is to ensure that a citizen who enters into a consumer contract can readily determine what his or her rights and obligations are, the type of legislation which best meets those goals should be determined. (By "readily determine" we mean that the individual can read through the contract, pointing to and describing the rights and responsibilities that are essential in the agreement.) Thus policymakers need the answers to such questions as:

1. In general, has the plain language legislation actually aided the consumer? Are "simplified" contracts easier to use and understand? The answer will provide empirical evidence to help evaluate the criteria in the various plain language laws.

2. What characteristics of style and design provide the greatest ease of use and comprehension? That is, what features of language and of layout and design distinguish contracts that are easy to use from those that are difficult to use? The answer will provide guidance to writers and document designers as well as to evaluators of documents.[2]
3. What characterizes the document design strategies of the companies which produce the most usable and comprehensible contracts? What design strategies or organizational strategies distinguish the superb performers, from ordinary designers? The answer will give us a better understanding of the document design process, will provide guidance to document designers, and may help legislators develop effective and meaningful plain language laws.
4. What is the best model for plain language legislation? That is, what legislative requirements result in documents that are easy to use and understand with the least cost to industry? The answer will provide guidance to government policymakers.

These questions require some further discussion. One way of answering the first question would be to analyze the extent to which companies have attempted to comply with the laws in states with plain language legislation. That analysis would seek to answer two questions: "What percentage of companies within a state has attempted to meet the standards of that state's plain language laws?"; and "Does that proportion vary significantly from state to state?" That strategy, however, may not be appropriate for assessing this legislation for two reasons. First, it would require the difficult, perhaps impossible, task of identifying every version of each type of contract in each of the states so that the contracts analyzed would represent a true cross-section of those in use. Second, it ignores the more pressing question of whether or not the revised contracts are actually easier to read. Unless the answer to that question is "yes," the degree of compliance is unimportant.

Questions 2 and 4 also require some discussion. It might seem that once the characteristics of documents that are easiest to use have been identified, legislators could simply make those characteristics the criteria in their plain language laws. The assumption there is that document designers need objective standards to guide their simplification efforts. Indeed there is some evidence that even well-intentioned writers need specific guidelines to follow, in part because revisers are often so familiar with the material in the contract that they find it difficult to anticipate the problems that consumers may have. Even if writers can recognize certain problems, they may need guidelines to solve them.

The West Virginia plain language law itself is a good example of this difficulty. The law is designed to promote clear language but is itself hard to understand. It reads in part:

> A violation of the provisions of this section shall not render any agreement void or voidable: Provided, that if a consumer at the time of entering into a consumer transaction or anytime thereafter, requests of the other party

thereto that the agreement evidencing the consumer transaction be changed or written in a manner to conform with this section, and that request is refused, then a consumer shall have a cause of action to require a consumer agreement not in conformity with the provisions of this section to be reformed (West Virginia 1981).

No doubt the drafters intended the law to be as clear as possible, but they still had trouble producing a law that the average citizen could understand. Similarly, experts rewriting contracts and other documents may not know how to make the documents readable unless they have some clear guidelines to follow.

There are, however, two main drawbacks to precise standards. The first is that they are inflexible and cannot be adapted to suit different conditions. For example, when consumers are dealing with contracts they use often, they may be better able to understand longer sentences or paragraphs than when they are reading unfamiliar contracts. Strict numerical guidelines do not allow courts or document designers to take such factors into account.

The second drawback is that the criteria, while associated with consumers' ease of use, may not guarantee clear writing. Concentrating simply on satisfying the criteria may lead document designers to disregard, or even distort, the meaning of the text. Readability tests, which form the basis of these criteria, were designed to assess texts, not to guide writers in writing them. Thus document designers need to aim at presenting their meaning clearly as they write, and apply the readability formulas or other objective criteria only afterwards.

Recent research in the Navy confirms this opinion. Duffy and Kablance (1982) found that documents that had been revised in good faith, but only to satisfy a readability criterion, were no easier for people to use than the unrevised documents (See also Duffy 1985; and Duffy, Smith, and Post 1987).

In sum, then, legislators may not be able to ensure that contracts are easy to use by simply giving document designers specific standards to meet. Therefore, identifying the characteristics of good contracts, the goal of question 2, and developing a model for effective legislation, the goal of question 4, need to be addressed separately. Any evaluation, therefore, will require two phases: first, an analysis of contracts and consumers' ability to use them; and second, an analysis of the design process which led to the contracts that are easiest to use.

Analysis of contracts and consumers' ability to use them

Types of contracts. The first phase might focus on three classes of consumer contracts: consumer installment credit, health insurance, and bank loans. We suggest these three kinds of contracts because: they are ones which almost all adults in our society enter into at some time; the restrictions and conditions in the contracts may have very serious implications for the consumer; the contracts vary considerably in complexity and length; and there are significant differences in the conditions and restrictions within each type of contract which makes comparison shopping worthwhile. In addition, the three types of contracts represent three very different kinds of business: the small intrastate business involved in installment contracts; the large, but primarily intrastate, banking business; and the large, primarily interstate insurance company.

States. The comparison among states with plain language laws is especially important. Currently plain language laws range from a simple statement of a general principle, like the New York law, to extensive and explicit requirements, like the Connecticut law. Three of the states (Minnesota, Maine, and New Jersey) provide for optional administrative review of the contracts by the state attorney general. Thus, in those states, in contrast to New York, an official assessment of compliance can be made without litigation. This procedure is designed to minimize risks for businesses by letting them know whether or not contracts meet plain language standards before the contracts go into circulation. The differences in the laws will need to be addressed, therefore, by comparing contracts from states that represent the range of approaches to drafting plain language laws. We would suggest New York, Minnesota, New Jersey, and Connecticut.

Socioeconomic status and education. At least two groups of consumers should be used in testing comprehension: consumers from the lower economic levels who do not have a high school diploma, and consumers from the lower-middle economic level who have a high school diploma and perhaps some college education. The former group — the poor and poorly educated — are the primary intended beneficiaries of plain language legislation. They are least able to deal with the complexity of legal documents and the ones for whom the consequences of a broken contract are the greatest. Unfortunately, research on the functional literacy skills of many members of this group suggest that many documents cannot be adequately simplified for the average member of this group to understand them. For example, nationwide testing of the functional literacy skills (the ability to accomplish the ordinary reading requirements of life) showed that members of this socioeconomic group had difficulty answering questions using even basic consumer information like train schedules, information from health agencies, and directions for locating community resources (Carver 1974). Ultimately, therefore, some combination of literacy instruction for this group as well as legislation requiring document simplification may be necessary before plain language legislation will prove useful to disadvantaged consumers, the group most in need of such help.

Comprehension. To test consumers' comprehension of contracts, consumers should be asked to read the contracts and then locate and interpret specific information in them. The questions should follow the presentation of short scenarios (short cases) and should represent the kind of information consumers usually need to know about a contract. Experts from consumer affairs and legal services should help generate a set of scenarios for each type of contract. Evaluators should assess both the accuracy of the responses and the time required to arrive at an answer.

Document characteristics. Evaluators should analyze the characteristics of each of the contracts they examine. There are two reasons for this analysis: first, to determine which contracts meet the legal requirements in each of the states; and second, to determine which characteristics of language and visual design best distinguish contracts that are easy to understand from those that are difficult to understand. Some of the characteristics that should be considered, for example, are: the use of unfamiliar terms; the average number of syllables per word; the average number of words presented; the organization of information, including the use of headings; the use of examples; typeface and layout. Every contract in the sample should be rigorously evaluated according to all the guidelines given in the various plain language laws, including all the objective criteria given in the New Jersey and Connecticut laws, and by other features identified as important in the research literature on document design.

In addition, evaluators should obtain holistic scores for each of the contracts. (The Minnesota, Maine, and New Jersey laws provide for optional review of contracts by the attorney general). Such a review provides a holistic assessment of the contract, i.e., a rating of the contract's overall comprehensibility. Raters will be asked to give two holistic scores for each contract, one for the style (language and content), and another for the visual design.

In states which provide for a review, the attorney general could use either plain language experts to make the holistic judgment or simply use available personnel. To reflect these two different conditions, two different panels should make the holistic evaluations: a panel of experts in plain language and consumer affairs; and a group of college freshmen. Each panel should be given instructions on rating taken from the Minnesota law and provided with a range of sample documents. The rating scale should range from "very comprehensible" to "very difficult to comprehend," with the midpoint representing "legal acceptability" in the judgment of the panel.

The evaluators should then analyze all the assessments of the contracts (the holistic judgments and the scores according to the objective criteria), along with the results from the comprehension testing, using multi-dimensional scaling techniques to determine what features of the documents best characterize the most comprehensible contracts. The scores of each contract should then be aggregated as necessary to score the contract as complying or not complying with each of the plain language laws. Finally the evaluators should determine whether the contracts that meet the standards of a particular state were more comprehensible by actual consumer-testing than contracts below the standards.

Document samples. Only documents which have been revised to meet the plain language requirements should be sampled, perhaps 15 contracts of each type (installment, bank, and insurance), from each of the four states, for a total of 180 contracts.

Consumer sample. The entire sample of consumers can come from one geographic area since there is no need to match people to the states from which contracts were obtained. Demographics of the area should be used to identify geographic locales having the two socioeconomic groups of interest (see Socioeconomic status and education, above). A sample of 500 or more should provide useful results.

Analysis of design strategy for exemplary contracts

It may well be that the critical factor in the design of comprehensible contracts rests in the expertise of the document designer and the approach to the design task rather than in any characteristics of the contract. We would in fact predict this to be the case (see, for example, Duffy 1981; and Duffy, Curran, and Sass 1983). Thus we see the analysis of the design process as a critical component of the project: what do good designers have in common that distinguishes them from less skillful designers?

Identifying approximately five of the most comprehensible contracts of each contract type and conducting preliminary interviews with the designers of these contracts would be a useful first step in answering that question, as would conducting interviews with designers in several groups who developed contracts which received low scores in the comprehension testing. The interviews should determine first how the groups are managed and then how the groups are organized, the qualifications and duties of the members of the groups, the procedures they follow, etc.

A next step should be an analysis of the "problem-solving strategies" (Flower and Hayes 1977) of individual designers, that is, the way designers conceive of their task in writing a contract and the steps they take to carry out their intentions. Designing a contract is basically a problem-solving task, figuring out how to present particular information to the consumer. In analyzing the designers' problem-solving strategies, the researchers should pay particular attention to how designers formulate the problem, how that formulation changes as the task progresses, what constraints or obstacles the designers perceive, and what strategies they use to respond to these constraints. The best way to determine such matters is to ask the designers to think aloud while performing the design task. These "think aloud" protocols should be recorded, transcribed, and then analyzed to identify the problem-solving behavior. This protocol approach is widely used in the analysis of other problem-solving behavior like composing a text or solving engineering problems and has been found to be a rich source of data for understanding that behavior (see, for example, Hayes and Flower 1980; Ericsson and Simon 1984).

Such analyses should provide the bases for organizing the document-design process and for training designers in how to approach document design.

1. For the undergraduate course, see Dixie Goswami, Janice R. Redish, Daniel B. Felker, and Alan Siegel, *Writing in the Professions,* Washington, D.C.: American Institutes of Research, with Siegel and Gale (1981). Two graduate programs are available at Carnegie-Mellon University: Master of Arts in Professional Writing, and Ph.D. in Rhetoric.

2. The term "document design" is awkward, but it has been adopted widely to include both the verbal and visual aspects of documents.

REFERENCES

Malcolm Baldridge. "Plain English in Business and Government," U.S. Department of Commerce, *The Productivity of Plain English* (pamphlet), (Washington: United States Government Printing Office, 1983).

Jimmy Carter a. "Improving Government Regulations," Statement on Executive Order 12044, March 23, 1978, *Public Papers of the Presidents of the United States*. Book I (Washington: United States Government Printing Office, 1979).

Jimmy Carter b. Executive Order 12044, March 23, 1978.

Ronald Carver. "Improving Reading Comprehension: Measuring Readability" AD-780 448, NTIS (Washington: United States Department of Commerce, 1974).

Michael Clive and Francine Russo. "The Plain English Movement in America: A View from the Front," *Information Design Journal,* 2 (1981), pp. 208-214.

Connecticut Public Act No. 79-532.

Thomas M. Duffy. "Organizing and Utilizing Document Design Options," *Information Design Journal,* 2 (1981), pp. 256-266.

Thomas Duffy, Thomas Curran, and Del Sass. "Document Design for Technical Job Tasks: An Evaluation," *Human Factors,* 25, no. 2 (1983), pp. 143-160.

Thomas Duffy and Paula Kablance. "Testing a Readable Writing Approach to Text Revision," *Journal of Educational Psychology,* 74 (1982), pp. 733-748.

T. Duffy, G. Smith, and T. Post. "Technical Manual Production: An Examination of Four Systems," *Written Communication,* 4 (1987), pp. 370–93.

Thomas M. Duffy and Robert Waller. "Readability Formulas: What's the Use?" *Designing Usable Texts* (Orlando, FL: Academic Press, 1985).

K. Anders Ericsson and Herbert Simon. *Protocol Analysis: Verbal Reports as Data* (Cambridge, MA: MIT Press, Bradford Books, 1984).

Linda Flower and John R. Hayes. "Problem-Solving Strategies and the Writing Process," *College English,* 39 (1977), pp. 449-261.

John R. Hayes and Linda Flower. "Uncovering Cognitive Processes in Writing: An Introduction to Protocol Analysis" In *Research in Writing: Principles and Methods* P. Mosenthal, L. Tamor, and S. Walmsley (Eds.), (New York: Longman, 1980).

Gregory M. Jones. "Confessions of a Reg Writer," *The Washington Post,* October 18, 1980; reprinted in William M. Schutte and Erwin R. Steinberg, *Communications in Business and Industry* (New York: Holt, Rinehart and Winston, 1983), pp. 22-24.

New York General Obligations, Sec. 5-702 *(Mckinney's Consolidated Laws of New York Annotated,* Chapter 136, 1978).

Office of Consumer Affairs, United States Department of Commerce. *How Plain English Works for Business* (Washington: United States Government Printing Office, 1984).

Pennsylvania Citizens' Consumer Council. *Statement of the Pennsylvania Citizens' Consumer Council in Support of House Bill 538 –" The Plain Language Consumer Contract Act,"* April 28, 1983.

Ronald Reagan. Executive Order 12290, February 23, 1981, *Compilation of Presidential Document,* February 23, 1981, Vol. 17 (Washington: United States Government Printing Office, 1981).

West Virginia Code. Sec. 46-A-109 (1981 Supp.).

The Plain English Movement in Australia and the United Kingdom

Robert D. Eagleson
University of Sydney

The current movement in favor of plain English has become firmly established in both Australia and the United Kingdom. Australian commercial enterprises quickly followed the example of the United States in adopting plain language, especially in insurance policies. The United Kingdom was slower but it was eventually to take a strong lead in the early 1980s within official circles, the British government developing highly organized policies and programs of action and instituting systematic processes of review. The movement in the United Kingdom has also been distinctive in having organized grassroots support.

A critical step in underpinning progress in plain English was taken when the Victorian government set up an investigation particularly of the drafting of legislation but also of other legal documents. The report Plain English and the Law *of the Victorian Law Reform Commission established that highly complex pieces of legislation could be written in plain English by redrafting a major Act of Parliament. This dismantles the central bastion of gobbledegook.*

Continued backing for the movement has risen as a result of the impressive savings that plain English policies have produced. Both commercial and government enterprises in both countries have reported extensive economic benefits, so much so that writing divisions are now being seen as important components in improving productivity and efficiency within organizations. The result has been to turn the spotlight on the inadequacies in the training of writers while at the same time vigorously maintaining the rights of the audience for comprehensible documents.

In 1976 Australia joined the movement for plain English when NRMA Insurance—the insurance subsidiary of the National Road Motorists Association of New South Wales—issued its first *Plain English Car Insurance Policy.* During a visit to the United States in the previous year the senior executive of the company had witnessed the introduction of plain English policies there. In 1974 in response to the consumer movement, the Sentry Life Insurance Company had undertaken a survey of its customers to discover what they wanted in insurance policies. Somewhat to the surprise of the company, a desire for comprehensible documents was among the more frequent responses. As a result Sentry rewrote its policies and was one of the first companies in the world to release a plain English insurance policy in 1975.

The Sentry policy was very much a response to customer wishes, a recognition of the need and right of an audience to understand documents it is expected to read and comply with. The NRMA venture—which was the first in Australia—was essentially a marketing exercise, a means of winning more business for the company. The success depended on the policy reaching a high standard of plainness, and as a result the social benefits for the public in readable documents were thoroughly covered, though these came as a consequence and were the means rather than the motivating force.

It was not long before the policy began to reveal another side to plain English. The consumer movement had stressed the rights of the public. Once in operation, plain English documents showed that there were considerable economic benefits for the procedures in terms of improved efficiency and productivity, that they were in fact good business far more than had been imagined. The NRMA discovered, for instance, that the number of invalid claims fell dramatically. Where previously customers had submitted accident claims and had left the responsibility to the company to work out whether the policy applied or not, now that the policy was comprehensible they accepted more of the responsibility themselves. The company also found that it took much less time to train staff, that staff gave more accurate advice to customers, and that junior staff had less need to interrupt senior staff for interpretations of various sections in the policies. Most pleasing of all, litigation declined, confirming that plain English could work successfully in law. These advantages together with the appreciative acceptance by the public encouraged the company to convert all its policies to plain English to the great gain of the public. Plain English drew writers and readers together, solving the different problems of each simultaneously.

The next step toward plain English in Australia by the Real Estate Institute of NSW also had the "sellers" rather than the customers uppermost in mind. Because the state government was threatening to make changes to tenancy laws, the landlords felt they should move first to show themselves in a more favorable light and possibly to stave off adverse regulations by producing their own improved standard agreement for residential tenancies. Despite this more narrow objective, the exercise uncovered another important benefit for the public in having documents in plain English. Practices whose inequity in favor of one party was concealed by gobbledegook were exposed in their blatant one-sidedness when expressed in plain English. Even the more hardened landlords were not prepared to have some conditions in their original form. They agreed to remove the more obvious imbalances and in the process the tenants benefited, thus substantiating that the role of plain English in protecting parties in an agreement goes beyond the removal of incomprehensibility.

Other insurance companies soon followed the lead of the NRMA, though not all, and some still have not adopted plain English in Australia. Several banks and finance companies also began to improve their documents, but progress was piecemeal, and restricted largely to the private sector.

In the United Kingdom, the movement came a little later. Again insurance led the way, with the first plain English insurance policy produced by General Accident Fire and Life Assurance Corporation appearing in 1979 three years after the first Australian venture, and five years after the commencement in the United States.

A GRASSROOTS MOVEMENT

Far more significant in the United Kingdom was the launching also in 1979 of the Plain English Campaign. This was a grassroots movement initiated by Martin Cutts and Chrissie Maher (1986:10). They had been working through their private consultancy firm with one or two government departments and local authorities trying to improve their forms but felt they were making little headway. They took the lead in establishing the campaign to obtain a broader initiative and as a means of rallying forces in the community to obtain change more rapidly.

Spurred no doubt by its efforts, the campaign soon found a strong ally in and important support from the National Consumer Council, a government financed body. In 1980 the council commissioned Vernon's *Gobbledegook* which was both an exposé of the senselessness of convoluted language and clumsy design in official forms and a signpost toward solutions.

Perhaps the most effective activity of the Plain English Campaign is its annual awards competition. Organizations are invited to enter their best forms, leaflets, or agreements for assessment by an independent panel of judges. The public is invited to send in examples of gobbledegook which have troubled them during the year. The awards—trophies for the winners, novel booby prizes for the perpetrators of gobbledegook—are presented at a very public ceremony which always receives extensive press, radio, and television coverage. The embarrassment caused to private and government organizations by receiving a booby prize has been a powerful incentive to improve.

The campaign is a community grouping outside government, industry, and commerce. While its membership is small and while Cutts and Maher remain the leading spirits, it is nonetheless a distinctly British contribution to the plain English movement. No such body exists in Australia. While consumer associations from time to time may criticize individual forms, there is no systematic program. As a result the major impetus always has to come from the inside. The movement lacks the impulse that a public watchdog can give, and which the Plain English Campaign provides so visibly in the United Kingdom.

STIRRINGS IN GOVERNMENT

In 1980 the Australian government set up the Task Force on Departmental Information. The objective was purely economic: the government wanted to improve the efficiency of official publishing and advertising services. But in the course of its investigations, the task force received complaints about the obscurity of government documents and forms. Its own consultant on the readability of Australian government publications noted that "many within the target audiences for government information would find much of the information incomprehensible" (*Report* 1982:xviii) and the task force itself observed "in general there was widespread indifference to the needs of the disadvantaged." A few departments responded to the recommendations of the report of the task force for more comprehensible documents, but the action was spasmodic and individual. The most notable example was the Department of Social Security which showed a commendable concern for pensioners and the unemployed, placing the emphasis for its activity on helping the disadvantaged.

Far more substantial developments were occurring in the United Kingdom at the same time. Having demonstrated the efficiences and economies that could be obtained through more effective forms and documents at Marks and Spenser, a leading British clothing retailer, Sir Derek Raynor was commissioned by the British government to survey the operations of the British bureaucracy. His investigations led quickly to the formulation of the white paper *Administrate Forms in Government* (Cmnd.8504) which was released in February 1982. It was in effect the declaration of a plain English policy for official publications by the British government.

A WHITE PAPER ON PLAIN ENGLISH

The commanding purpose of the policy was stated in the white paper to be good service to the public, entailing the public's entitlement to understand government documents so that it can know its rights and duties and avail itself of benefits offered in government programs. The paper recognized that documents were an important means of communication between government and public, and indeed for many people "the face and voice" of government. Hence it argued for the great need for them to be written in plain and comprehensible language.

The white paper also made it clear that the policy was to be cost-effective. It was to make savings for the public: for example, a well-designed form should be easier to complete and less time-consuming. It was also to lead to large savings for the government: the processing of forms should be simpler and more efficient, the rate of error in completion and the number of queries should be lower.

That these economic benefits for government were the compelling pressure for the formulation of the policy does not mean that the social benefits for the public in terms of more comprehensible documents were lessened or ignored. On the contrary attention to them was probably strengthened and protected by these ties, for organizations are far more likely to change to plain English and to continue their commitment to it if they can see real gains in it for themselves. That there were other advantages in addition to the social benefits only lent more robustness to the British government's resolve to introduce plain English.

A systematic program

There have been important advantages from the program emerging in such a way as the result of a white paper. The goals for departments were clearly delineated from the start, and the basic principles firmly set. It has led to the British tackling the problem of gobbledegook in official circles in a systematic way.

The British government recognized that it was easy for such programs to lose their momentum and to lag. To avoid this, it appointed the Management and Personnel Office as an agency to provide constant stimulus from the center and to act as a watchdog. The office established a Central Forms Unit to undertake these roles by coordinating work within departments, identifying and advertising good practice, providing suitable training courses, and reporting on progress.

Action within departments

So that the implementation of the policy was not left to chance, departments articulated comprehensive programs of action and established definite lines of responsibility, requiring a commitment from all levels of staff.

In most departments, steering committees—known as either Form Review Groups or Forms and Notices Review Committees—were established. These consist of senior members of management who may have no direct involvement themselves in the preparation of documents, but who are charged with overseeing progress within their particular departments. The committees have power to review any document and to call for improvement. The very seniority of the membership lends weight to the authority they have been given to promote review and reform.

To give practical effect to the goals of the policy most departments either strengthened any existing policies or established forms design units. These are specialist units composed of staff with expertise in various aspects of the preparation of documents, some with training in linguistics, others with a background in typography and layout, others with a wide knowledge of the operation of the department itself.

The primary function of the forms design unit is to give direction and to exercise quality control within a department. While it is given responsibility for specific documents, notably those with the largest and most general distribution such as the Tax Return and the Supplementary Benefit Claim Form, this part of its work is carried out as much in the role of a pacesetter as a producer. By its example, which ensures that the unit has a practical involvement in the work of a department and in the preparation of documents, the unit can set the pattern for amendments elsewhere and demonstrate to other sections how specific changes can be made in the context of its department's activities.

Equally important, on the basis of the expertise of its members and their experience in producing and testing documents, a departmental forms design unit prepares guidelines for other sections which are producing forms and leaflets within the department. Some of those which have been produced already, such as the "Good Forms Guide" in the Department of Health and Social Security, are quite extensive, while others are limited in scope to particular matters, such as the short thesaurus on difficult words and their plain English equivalent which was produced by the unit in Customs and Excise.

A forms design unit is also required to exercise strict control over all new and revised forms emanating in its department. All requests for new forms and for reprints must first be submitted to the unit for vetting and for approval of language and design. This authority is vested in the unit by clear statements in departmental procedures and is seen as an important means of improving quality and preventing the proliferation of unnecessary documents.

Precise monitoring of review

As a means of enabling fixed programs of review to be formulated, departments have compiled an inventory of all documents in use. Within the inventory a set of criteria is applied to classify the documents according to their status or degree of importance, such as size of circulation, importance or sensitivity of the policy involved, error rate in returned forms, internal or external use, cost of production, completion and processing. This further step of classification has provided a basis for establishing priorities in conducting reviews, enabling efforts in revision to be related to benefits of improvement.

Most departments have now set up programs that are directed to obtaining systematic reviews of all forms every three to four years. In the process they are changing past policy from a demand-led approach in which a form was looked at only when it had to be reprinted—and hence under pressure of time—to undertaking a regular rolling program of review.

Backing for the U.K. program

A significant element in the speedy implementation of the policy in the United Kingdom has been the quality and the stature of the backing it has received. The prime minister was a main source of initiative behind the policy and continued to lend weight to its promotion both in public statements and in attendance at functions where successful revision of documents were displayed. Moreover as part of the policy all government departments and instrumentalities are required to report on progress in the improvement of their documents through their ministers to the prime minister each year. The backing of the highest authority thus remains very visible.

This obvious support from the prime minister, and especially the requirement of an annual report have led to ministers and permanent heads of departments becoming more actively involved even where previously they may not have had an overt commitment to the policy. This in turn, coupled with pressure from the Plain English Campaign competitions, has helped to reduce obstacles for advocates and supporters of the program within departments and has ensured steady progress with the policy.

Impressive U.K. results

While not all departments have responded as energetically as others, nevertheless the results of the British policy are impressive. By the end of the third year, over 65,000 forms had been reviewed, nearly 16,000 had been eliminated as unnecessary, and more than 21,000 had been revised, most of them extensively. Equally to the point, as far as the government's goals are concerned, huge cost savings have been achieved. The Customs and Excise Department reduced the error rate on one form from 55% to 3%, reducing the processing time alone by 3,700 hours. The Department of Defense revised a claim form for traveling allowances, reducing the time spent in completing it by 10%, the processing time by 15% and the error rate by 50%. The savings amount to £400,000 a year for a one-off outlay for redesign costs of £12,000. The Department of Health and Social Security has a string of successes, reaching over £1 million a year on some forms. The third annual report for 1984–85 to the British government alone identified fresh savings of over £4 million a year. The government now feels that the program is securely established so that it is cutting back on the need for regular annual reporting for a trial period.

OFFICIAL ACTION IN AUSTRALIA

Although the report of the Task Force on Departmental Information and the efforts of private industry had prompted the Department of Social Security in Australia to seek to improve the comprehensibility of its forms, it was the development in government circles in the United Kingdom that gave the greater spur for change. This provided a basis and a model for recommendations to the Australian government, which in February 1984 declared a policy for plain English in official documents.

The Australian program has not been as systematic as the British one. In the first place, the drive for it came more from within the public service itself than from the government. Secondly, government backing, while real, has never been as pointed as in the United Kingdom, and has been distracted by cabinet reshuffles and elections. The Information Co-ordination Branch, which was given the responsibility for coordinating the policy, has been variously in the Departments of the Special Minister of State, Sport, Recreation and Tourism, and now Administrative Services. Despite these discontinuities, there have been important and publicly visible advances at the federal level of government. The three departments with extensive contact with the public—Health, Social Security, and Taxation—have completely overhauled their most frequently used forms and leaflets and have brought them to a high standard of plain English, so much so that the Taxation Office has received letters of praise from the public for the first time in its history. Social Security has also led departments in the critical process of rewriting standard letters, recognizing that it communicates more often and more extensively with the public through an array of these letters than through any other publication. Its endeavours here have been more than justified by the improvements in the day-to-day operation of its various schemes and its greater success rate in its contact with members of the public. Other departments, or sections within departments, have adopted plain English, but its spread has been more as a result of the example and achievement of others than by pointed directive.

The Information Co-ordination Branch appointed an outside consultant as Special Adviser on Plain English to help develop the program and to conduct workshops within departments to train selected officers in preparing documents in plain English. It is now producing a manual, "Writing in Plain English," which is designed to be used both by individuals and by training sections within departments as a basis for constructing more extensive and continuing training programs for staff.

These activities at the federal level and within private industry are also having an influence on state and local government agencies. More and more of these are turning to plain English as its feasibility and superiority in communication has been substantiated, and as its advantages for organizations have become persuasive.

TESTING

In both countries, as elsewhere, testing has been recognized as an essential component of a plain English program. It has been undertaken at all stages in the development of a document, from the initial planning of policy to the final draft and on to the finished version in use. It is largely concerned with the reactions of the intended audience, and so the needs of the various segments of the community receive due attention even though other objectives may be primary for the initiation of the document. But testing is also regularly conducted on the operation of a form, for instance, within an organization with the result that the contribution of plain English in increasing efficiency has been continually substantiated. This in turn has underpinned the commitment of organizations to plain English.

Testing not only has had immediate benefits for the document in hand but has also pointed the way to further research into matters with wider applications. More recent examples include the study by the UK Department of Health and Social Security (1984, 1986) into the effectiveness of forms and into methods of determining which forms are more likely to repay extensive rewriting and redesign. Inland Revenue has promoted a study of taxation and communication (James and others [1985]) and a review of international approaches (Foers [1987]). In Australia, the Information Co-ordination Branch initiated an investigation of appropriate spaces for names and addresses on forms (Sless [1987]).

PLAIN ENGLISH IN LEGISLATION

While the early Australian ventures in plain English were largely imitative or influenced by what was happening elsewhere, in 1985 one state government took an important initiative. The Victorian Attorney-General, James Kennan, gave the Victorian Law Reform Commission a reference "to enquire into and review current techniques, principles and practices of drafting legislation, legal agreements and those Government forms which affect legal rights and obligations, in order to recommend what steps should be taken to adopt a plain English drafting style."

Because of the overbearing influence the language of legislation exerts, with lawyers and officials hesitant to depart too far from the wording of acts and subordinate legislation even in correspondence, many in Australia and overseas have longed for such a development. There have been studies and investigations to bring about improvements, such as the Renton Committee in the United Kingdom in 1975, but these have never yielded startling changes. There are also the efforts of the Office of the Law Revision Counsel in the Congress itself, and for example, of the Revisions of Statutes in Maryland and Minnesota, but these are concerned with reworking legislation once it has been passed and as part of the codification process. The purpose of the Victorian study was to have legislation written in plain English from the start.

The possibility of initial drafting in plain English had been established by a pilot project on a Coroners Act in 1985, which had also revealed how much needed to be done. Activities in 1986 were devoted to examining the language, design and stages in development of acts and other legal documents, and in producing models. The outcome was a final report tabled in the Victorian Parliament in October 1987 which affirmed the viability of plain English for legislative drafting.

Perhaps the most significant part of the Law Reform Commission's report is appendix 2 which contains a rewritten version of the Companies (Acquisition of Shares) Act—the Takeovers Code—alongside the original version and together with a commentary on the changes. The code is generally acknowledged as an extremely complex piece of legislation. It was chosen deliberately to answer charges that plain English can cope only with simple, legal documents, and to demonstrate that plain English can work over an extensive document (the original version of the code runs to eighty printed pages). A team of private legal practitioners expert in takeovers have attested to the accuracy and effectiveness of the plain English version.

The demonstration model—and indeed the recommendations in the report itself—is not concerned only with narrow matters of language such as sentence structure and vocabulary choice. It takes up matters of larger discourse structure and introduced changes in the presentation of material. Housekeeping matters, such as provisions dealing with commencement dates, jurisdiction, and definitions, are moved to the end rather than appearing first as in the traditional style, in recognition of the fact that most readers are interested in the substantive matters and not these ancillary issues. Defined terms are clearly marked each time they appear, so that readers are given warning to check the definitions section, and attention is given to layout.

The reference to the Commission includes general legal documents as well as legislation, and appendix 3 contains rewritten versions of a common mortgage, loan agreement, and a summons. The summons had been developed in 1986 and will lead to the equivalent of a saving of at least $400,000 a year by allowing some thirty members of staff to be redeployed from outmoded clerical activities to more productive work.

The Victorian government has already moved to implement a number of the report's recommendations and to direct parliamentary counsel to adopt the principles of plain English drafting as set out in the report and in the manual of drafting which constitutes appendix 1. The government is also seeking to establish a National Institute of Legal Drafting at Monash University (Melbourne) to offer postgraduate training in drafting generally and especially for intending legislative drafters, and to undertake research in the use of plain English in legal documents.

The ramifications of this Victorian enterprise are far-reaching. It means that the area of writing that has been the most resistant to change and that has had a baleful influence on other types of official writing has been mastered. There are no longer any rational grounds for clinging to traditional models.

PLAIN ENGLISH LAWS

Since 1984 the National Consumer Council and the Plain English Campaign in the United Kingdom have been advocating a plain English law. A draft bill has been prepared and there have been attempts to have it introduced in Parliament by private members. So far the government has not been receptive. The proposed bill is modeled on corresponding American legislation.

While some acts require consumer documents to be written in a clear style and printed in legible typefaces, there is no real support for a general plain English law in Australia. There is a concern that such a law might stifle creativity, especially when so much more is still to be learned about plain English itself and how readers and writers function. It is felt better to leave the initiative with private and public authorities and to the pressures of competition.

WRITERS—THE NEW NEEDY

The central platform of the plain English movement has always been the right of readers to understand readily any document that affects them closely. This emphasis must be maintained. But as the movement has developed, and especially in the light of the experiences with legislation and legal documents, it has become increasingly apparent that we must broaden our concern to include the other party in the communicative act—the writers. While the audience may be oppressed by gobbledegook, writers are ensnared by it. They do not produce it so much by deliberate intent but because by imitation, training, and exposure they have come to feel it is the only way to write. They too are in desperate need of being released from the tyranny of gobbledegook.

Evidence of this need continually surfaces. During 1987 a lawyer in a major legal firm strenuously resisted the pleas of the NSW Department of Housing to rewrite a standard lease in plain English for the sake of pensioners on the grounds that plain English was not legally valid. And this in the state in which plain English had been introduced in Australia, and in which plain English insurance policies and other documents had been operating for some ten years without having their legal validity ever challenged in court. Worse, the lawyer wanted to persist with such drafting as:

3. To the full extent of the covenants next hereinafter shortly noted as the same are set forth in words at length in the Second column of Part 2 of the Fourth Schedule to the Conveyancing Act 1919 as amended;
 (1) that the Lessee covenants with the Lessor to pay rent;
 (6) and that the Lessor may enter and view state of repair and that the Lessee will repair according to notice in writing and that the Lessor may repair;
 (7) and that the Lessor may enter and carry out requirements of public authorities and repair under the Lease.

Nor did the lawyer seem constrained to make any changes at all in extracting clauses from other leases, even though there were glaring inconsistencies and clumsiness as 4 (c) and 4 (d) show:

4. (c) At his own expense with all due dispatch to make any repairs, additions, improvements or alterations and do all other things necessary to comply with any notice or order issued by any public authority relating to the premises and occasioned by the use of the premises by the Lessee or by the act, neglect or default of the Lessee, his employees, agents, invitees or licensees.

4. (d) To pay for and make good all damage which may be caused to the premises or the Building by reason of any water, gas, drainage or electrical installations, appliances or fittings being left open, obstructed, misused or damaged in any way by the Lessee its servants, agents, invitees or Contractors.

Moreover, other lawyers and senior officers in the department had passed the lease and not one of them had recognized that clause 3 for instance, lacks a grammatical subject and is nonsense. There is something radically wrong when professionals produce clauses like these, or sentences which run for three hundred or more words even though they know from their own experiences as readers that sentences of inordinate length are difficult to comprehend.

Plain English is not an innovation of the last decade. It has been practiced in the past and there have been pleas for it in earlier centuries, for example by Bentham and Coode, and even more recently by Flesch in the 1940s and Gowers in the 1950s. If this new movement is not to flounder, if we are not going to be faced with having to resuscitate plain English continually, then thoroughgoing attention must be given to the needs of adult writers to help them eradicate deep-seated misconceptions about language and to develop a congenial atmosphere and context in which they can feel free to practice plain English. (See also Dykstra [1987] who has expressed similar views on the needs of writers in Canada.) It is being recognized more clearly that this other side to the coin of plain English must be tackled as vigorously as the rights of the audience. It is this perception which lies largely behind the proposals for an institute of drafting in the Victorian Law Reform Commission's report.

In both Australia and the United Kingdom plain English has proved that it provides a more effective mode of communication, enabling all parties to understand without leaving one party with the power of interpretation. It has also established its worth over traditional modes as a more efficient means of giving and getting information. What remains to be done is to encourage the widespread adoption of plain English and to secure its persistence through the education of writers.

REFERENCES

Australia Dept. of the Prime Minister & Cabinet (1980), *Report of Task Force on Department Information* (Canberra: Australian Government Publishing Service).

Bentham, J. (1838) "Rationale of Judicial Evidence" in Bowring, J. (ed.), *Works of Jeremy Bentham* (Edinburgh: William Tate) Vol. II.

Coode, G. (1845) *Legislative Expression* (London: HMSO).

Cutts, M. and Maher, C. (1986) *The Plain English Story* (Stockport, U.K.: Plain English Campaign).

Dykstra, G. (1987) "Plain Language, Legal Documents and Forms." Paper presented at Canadian Institute for the Administration of Justice Seminar on Legislative Drafting and Interpretation, Ottawa, August 19–21.

Eagleson, R. D. (1983) *The Reform of Documents within Government Departments in the United Kingdom* (Canberra: Information Co-ordination Branch, Department of the Special Minister of State).

Eagleson, R. D. (1985) *Progress in Plain English* (Canberra: Information Co-ordination Branch, Department of Sport, Recreation and Tourism).

Eagleson, R. D. See Victorian Law Reform Commission (1986), (1987).

Flesch, R. (1979) *How to Write Plain English* (New York: Harper & Row).

Foers, J. M. (1987) *Forms Design: An International Perspective* (London: Inland Revenue).

Gowers, E. (1954) *The Complete Plain Words* (London: HMSO). 2d ed. 1973, ed. by Fraser, B; 3d ed. 1986, ed. by Greenbaum, S. and Whitcut, J.

James, S., Lewis, A., and Allison, F. (1985) The Comprehensibility of Taxation and Communications. Mimeo (in press).

Renton Committee (1975) *The Preparation of Legislation* (London: Sweet & Maxwell).

Sless, D. (1987) *Name and Address Please* (Canberra: Information Co-ordination Branch, Department of Sport, Recreation and Tourism).

U.K. Department of Health and Social Security (1984) *Forms Effectiveness Study* (London: DHSS). Research conducted by Coopers & Lybrand Associates.

U.K. Department of Health & Social Security (1986) *Effectiveness of Forms: Phase II* (London: DHSS). Research conducted by Logica Consultancy Ltd.

U.K. National Consumer Council (1984) *Plain Words for Consumers: the Case for a Plain Language Law* (London: NCC).

Vernon, T. (1980) *Gobbledegook* (London: National Consumer Council).
Victorian Law Reform Commission (1987) *Plain English and the Law* (Melbourne: VLRC).
Victorian Law Reform Commission (1986) *Legislation, Legal Rights and Plain English* (Melbourne: VLRC). Discussion Paper No. 1.

A Plain Language Centre for Canada

Gail Dykstra
Canadian Law Information Council, Toronto

Plain language legal documents are essential if Canada is to fulfill its goal of improving public access to the law. The role of plain language in achieving public accessibility and its uses in consumer contracts, insurance documents, and forms is explored in this paper. An explanation of the status of plain language laws is provided, along with an outline of the arguments for and against the use of plain language in statute drafting. The goals and activities of the Canadian Law Information Council's new Plain Language Centre are outlined.

INFORMATION

The Canadian Law Information Council (CLIC) is going to create a Plain Language Centre for Canada. After considering the present use of plain language in Canada and the experience of the Commonwealth and of the United States, CLIC has identified the critical missing element that impedes the acceptance of plain language in Canada. We need a plain language catalyst, a well-defined program which will fill the identified need for greater awareness of plain language in Canada; an information-resource service on plain language activities and materials; and a consulting service capable of providing research, editing, design and training to those preparing legal documents and forms.

The Plain Language Centre is intended to meet these needs by

- creating an awareness of plain language among the public and those who prepare legal and administrative documents and forms, and
- providing the tools, information, advice, training, and research needed to use the plain language approach to communicating information on the law.

The long-term objective is to create a self-sustaining CLIC Plain Language Centre to provide consulting services to governments, businesses, consumer groups, unions, the legal profession, and community and public legal education (PLE) groups.

Although specific activities and service priorities will be recommended by an Advisory Panel, CLIC has identified basic project directions and calculated an approximate cost to achieve these objectives. The Centre would be product- and service-oriented, offering a variety of practical solutions to the problems faced by those who want to use plain language in the legal profession, government, business, PLE, and consumer groups. During the three years of the program, CLIC will evaluate its activities and determine if they have met their objectives. CLIC will offer its services on a bilingual and bijuridical basis.

Each demonstration project will act as one of the building blocks to secure a foundation for the establishment of an ongoing Centre. During the pilot project the Advisory Panel and CLIC will work together to find contract and service funding to make the Centre self-sustaining in year four.

CLIC has received funding for this program from the Donner Canadian Foundation, which has dedicated $175,000 over a period of three years. The Law Foundation of Ontario has granted CLIC $50,000 for the Centre with the understanding CLIC can apply for additional funding in the coming years. CLIC has requested funds from other foundations, governments, and private companies in order to provide a sound financial base for the CLIC Plain Language Centre.

WHAT WILL CLIC DO?

The priorities for the CLIC plain language program are to work on consumer-oriented legal documents and administrative forms. These widely used documents must be made easier to use and easier to understand. It goes without saying that explanations of the law and government services for the public must be available in plain language. CLIC will

- teach those who prepare legal documents and administrative forms how to use plain language through "How to" workshops and training materials that we will create;
- let people use the experience of others in solving their own communication problems by providing information services and a comprehensive resource collection of plain language samples, instruction manuals, and descriptions of current plain language activities;
- create a CLIC plain language contract agency.

Training: "How to" workshops

In order to make sure our plain language program meets the needs of those who prepare legal documents, CLIC will undertake an extensive analysis of how lawyers are trained to write, what types of instruction would be most helpful, and when in the legal education process it would be most appropriate to provide plain language instructional modules, or full courses, for use in law schools and continuing legal education programs.

We will be doing needs assessments on the training needs of government officials and producers of public legal education materials in order to create "How to" workshops on writing and form design to meet their needs. CLIC is prepared to invest the time and energy needed to undertake a substantial research phase prior to designing "How to" courses or creating instructional guides. Training, if it's to be used six weeks after the course is given, must solve real life problems. In plain language terms, CLIC needs to find solutions to the common writing problems that occur every day. We will be calling on those who face these problems to help us find solutions. Professors of law, government form designers, lawyers and public legal educators from both common and civil law jurisdictions will be asked to help us in our research and to assist in the development and testing of "How to" courses.

CLIC has developed a list of plain language training products (table 1) that will accomplish our goals. Their research/development and marketing phases are outlined.

Samples and information services

When the Bank of Nova Scotia and the Royal Insurance Company were redrafting their consumer contracts, they worked in isolation. Neither was aware of the other's work even though they were in the same city, in fact, within a few blocks of one another. Although that was in 1978 and 1979, isolation is a continuing problem.

Those who have decided to use plain language want to learn from each other's experience—and can. Seeing how someone designed a tax form, for example, can help someone else redesign another. Seeing a plain language residential lease and hearing about the process used to create it can help someone shrink a four-page, single-spaced lease to a more manageable size.

Precedents

A comprehensive bank of samples of plain language legal documents and administrative forms is urgently required. Being able to use other people's documents to guide one through the plain language drafting process will save time and money and will increase the likelihood that one's new plain language document will be a success.

Gaps

CLIC will survey the existing plain language samples and determine where there are gaps. New samples of documents and forms will be created to fill these gaps and to add to our knowledge of how the plain language process works. It is essential that CLIC develop sufficient expertise to help the public and private sectors accurately predict what staff resources, time, and money will be required to initiate a plain language redo of their documents and forms.

Communication

Another solution to the problem of isolation will be the communication tools CLIC will create to put people in touch with others interested in plain language. Newsletters and a database of information on current activities, publications and conferences will connect people with plain language interests and experience.

CLIC will be able to maintain a comprehensive view of the plain language experience in Canada and monitor the areas that need plain language attention. Calling on the expertise of the CLIC plain language advisory board and experts in the field, CLIC will be able to effectively target its research efforts

Table 1. Products: "How to" Workshops

Product	*Research and Development Phase*	*Marketing Phase*
Law School Module for Legal Writing Courses • demonstration program developed and tested at selected law schools 1 English (Common Law) course 1 French (Civil Law) course	Set goals and objectives and explore potential ways to accomplish project Research training needs, current resources, potential solutions Consult with professionals in field Develop materials and create new materials as needed Test training program and materials Rework program and materials	Newsletter Information Sheets
Continuing Legal Education Workshops and Course Module —workshop for practicing lawyers —course module for use in subject-oriented CLE courses • demonstration workshop developed and tested in selected province • demonstration module developed and tested with selected CLE program	Set goals and objectives and explore potential ways to accomplish project Research training needs, current resources, potential solutions Consult with professionals in field Develop materials and create new materials as needed Test training program and materials Rework program and materials	Newsletter Articles and Columns Information Sheets Media Campaign
Public Legal Education Materials Workshop • demonstration workshop developed and tested with PLE organization in selected province	Set goals and objectives and explore potential ways to accomplish project Research training needs, current resources, potential solutions Consult with professionals in field Develop materials and create new materials as needed Test training program and materials Rework program and materials	Newsletter Articles and Columns Information Sheets
Government Communications Workshop on Three Areas of Plain Language —forms —legal documents —information on the law • demonstration products developed and tested	Set goals and objectives and explore potential ways to accomplish project Research training needs, current resources, potential solutions Consult with professionals in field Develop materials and create new materials as needed Test training program and materials Rework program and materials	Newsletter Articles and Columns Information Sheets
Self-Instructional Materials on Plain Language —printed materials —computer software • demonstration products developed and tested	Set goals and objectives and explore potential ways to accomplish project Research training needs, current resources, potential solutions Consult with professionals in field Develop materials and create new materials as needed Test training program and materials Rework program and materials	Newsletter Articles and Columns Information Sheets

(*continued*)

Table 1. (*Continued*)

Product	*Research and Development Phase*	*Marketing Phase*
Legal Writing Teachers' Workshop on Plain Language • demonstration workshop developed and tested with selected schools	Set goals and objectives and explore potential ways to accomplish project Research training needs, current resources, potential solutions Consult with professionals in field Develop materials and create new materials as needed Test training program and materials Rework program and materials	Newsletter

and produce practical recommendations for future action. Therefore, CLIC has developed a list of plain language information services and communication vehicles that will accomplish our goal. They are listed in table 2 along with the research/development and marketing phases that are essential for their implementation.

Consultation services

Plain language documents and forms save money. Companies and governments in the United States and the United Kingdom have found they can save thousands and sometimes millions of dollars every year by rewriting their forms and documents. In the United Kingdom, for example, the Department of Health and Social Security introduced plain language application forms for legal aid in 1984. It cost the department about $50,055 to develop and test the forms. They found they saved $2,917,290 in staff time every year by using the plain language forms. Canada can draw upon the experience of these countries and create a plain language consultation service to serve our needs.

Ever since CLIC started researching plain language, the council has been asked to undertake fee-for-service projects. We will respond to this growing need by investigating the possibility of providing a range of plain language consultation services. CLIC will then establish an agency that will be able to tackle small or large plain language redesign/rewrite projects.

The Plain Language Centre will be able to provide a full range of services.

Training

- "How to" workshops and self-instructional materials for government departments, individual companies, and law firms.
- Training for managers and lawyers who must supervise plain language projects.

Consulting Services

- Development of plain language forms and documents.
- Organization of in-house plain language projects.

Table 2. Products: Samples and Information Services

Product	*Research and Development Phase*	*Marketing Phase*
Advisory Service to give information on —current activities in Canada and elsewhere —articles and books on plain language —names of organizations and people working in the field • requests for information handled free of charge • bilingual/bijuridical services	Identify, acquire and organize samples Create samples to fill gaps Create and maintain files on people and activities in Canada and other jurisdictions	Newsletter Articles and Columns
"Samples Bank" of examples of previous plain language documents and forms • materials loaned on request • bilingual/bijuridical services	Identify, acquire and organize samples Create samples to fill gaps Create and maintain files on people and activities in Canada and other jurisdictions	Newsletter Articles and Columns Awards Media Campaign
Plain Language "Courtwatching Service" monitors judicial review of legal documents and forms. • series of articles prepared • bilingual/bijuridical services	Research past and current court cases with plain language implications	Newsletter Media Campaign
"Tip Sheets" on plain language solutions to communication problems • 4 sheets per year over two years • bilingual/bijuridical services	Research and prepare plain language instructions for solving communication problems	Newsletter
Information tools on plain language —CLIC Plain Language Newsletter • quarterly, free, bilingual/bijuridical —Columns and articles on plain language • 16 columns or articles made available to government and organizations' newsletters, bilingual —Booklet on plain language • 6–8 pages, camera-ready copy to be distributed by public and private sectors, bilingual	Research and produce newsletter, booklet and columns on plain language	Newsletter Articles and Columns Media Campaign
Speakers service on plain language • Respond on demand if expenses covered for travel • Bilingual	Research plain language issues	Media Campaign

CLIC will also be able to help governments and business assess their need for plain language documents and provide guidance on selecting communication consultants and how best to introduce plain language documents to their users/clients. CLIC has developed a list of potential plain language consultation services that will accomplish our goal. They are listed in table 3 along with the research/development and marketing phases that are essential for their implementation.

NATIONAL ADVISORY PANEL

CLIC believes it essential to involve representatives from the plain language target groups in the determination and implementation of the projects. Their ideas will help CLIC be more effective in developing and delivering good programs that help people know and use plain language. Their support will be critical in creating a permanent CLIC Plain Language Centre.

Table 3. Products: Consultation Services

Product	*Research and Development Phase*	*Marketing Phase*
Consulting Service: In-depth advisory services available to public and private sectors —needs assessments —arranging for plain language communications consultants' bids —matching client with communications company —overseeing contract —providing in-depth research assistance —advisory assistance throughout contract	CLIC's "How to" Workshops CLIC's Samples and Information Services	Newsletter Articles and Columns Media Campaign Awards Speakers Service
Training Workshops: Putting on plain language training workshops using CLIC's workshop approach	CLIC's "How to" Workshops CLIC's Samples and Information Services	Newsletter Articles and Columns Media Campaign Awards Speakers Service
Feasibility Study for a Plain Language Agency: Full service communication consulting company specializing in plain language documents and forms	Assess current Canadian resources Evaluate potential market Discover start-up and short term costs of setting up agency Determine appropriateness and feasibility of CLIC participation	Newsletter Articles and Columns Media Campaign Awards Speakers Service Individualized selling campaign Response to RFP's

The Plain Language Advisory Panel would consist of representatives from

- CLIC
- the legal profession, including representation from the judiciary, government, private practitioners and law schools
- the public, including representatives from consumers' groups, the PLEI network, the media, government communications, and the adult basic education/adult literacy movement.

The advisory panel will serve a function that goes beyond strict advice-giving. Because of the specialized nature of many of the Plain Language Centre's activities, CLIC recognizes the need for some panel members to serve as experts on the research teams responsible for "product-oriented" projects.

EVALUATION AND REPORTING

CLIC believes that an evaluation component is necessary to ensure the Plain Language Centre meets its goals and objectives and responds to evolving priorities throughout the three-year term of the pilot program. In this way a sound foundation will be created for the permanent Plain Language Centre. Drawing on its experience with the *PLE Evaluation Resourcebook,* CLIC has selected an evaluation process that will make maximum use of the personnel and financial resources and provide a continuing supply of information throughout the three-year term of the Law Foundation grant.

CLIC will develop a preliminary evaluation plan. This plan will identify program ad project goals and objectives, the personnel available to do the evaluation, and it will suggest timelines for the gathering of information and periodic analysis of that information. CLIC will then hire an evaluation consultant. Together, CLIC and the evaluator will create a practical evaluation framework and identify appropriate evaluation methods.

Technical Writing in Japan: Notes on a New Profession

Hiraku Amemiya
High Technology Communications, Tokyo

Until as recently as two or three years ago, technical writing received virtually no recognition as a profession in Japan. Most manuals were still written by engineers, and most people had never heard the term technical writer. This situation is changing, slowly but noticeably, primarily because of the wide dispersal of PCs and word processors among novice users. Software users have taken the lead, but in the past year or two some hardware manufacturers have established departments specifically for technical communication. Nevertheless, the technical writer's situation remains lamentable. There are virtually no education or training programs or even textbooks available to them, and they receive little or no understanding from management (a situation which translates into limited budgets and time). People who entered the technical communication field early (five years or so ago), especially, turned to the West for instruction. They learned a great deal from American models, in particular. But they soon found they had to adapt these lessons to the special needs and characteristics of Japanese users—with their different learning process, keyboard usage and expectations of customer-service personnel, for example. The result has been a user manual which differs in significant ways from its Western counterpart.

One day not too long ago I was startled by the sudden realization that technical writing had begun to make itself visible in Japan. Its presence remains somewhat shadowy, I admit, but the future of our profession looks much more promising now than it did just two or three years ago.

At that time most manuals were still written by engineers. It was not uncommon for U.S. dealers to tell Japanese computer companies that their hardware was excellent, their software okay, but their documentation just a joke. And this was not true only of most of the documentation for the export market; Japanese users suffered from the same jumbled manuals.

The Japanese computer manufacturers had a particularly strong conviction that good products would sell themselves. They also believed that if a product had a high potential, the customer should work hard to learn how to use it. Nor were beliefs like these altogether invalid. As long as the machines cost so much that only big customers could afford them, customers' complaints and questions could be handled by a limited number of field-support and customer-service personnel.

All this has changed. With the drastically lower cost of today's computer, ordinary people have been invited to participate in the computer age. Buying a PC at a computer shop is as easy as taking out a pizza from Shakey's. Most people, however, still have no clear notion of just what this computer age might mean to them. A refrigerator can make ice cubes, they tell themselves. A skillet is ideal for frying bacon. But what good can a computer do me? You say it can speak? Wow! But what's it saying, anyway?

Vacuum cleaners, bicycles, and other conventional single-function products can be sent out into the market without much explanation on the very safe premise that everybody will know just what they are and what to do with them. But multifunctional products such as computers, word processors, or cameras with dozens of automatic features don't allow users the luxury of fixed assumptions concerning their nature or use.

With the popularization of the computer in Japan, the need for professional technical writing became apparent. Little by little people began to realize how deep and dark was the gulf between the engineer and the user. The engineer's goals are to make the machine perform a certain set of actions and to determine what it should express on the output devices provided. The user, on the other hand, simply wants to get the job done as easily and quickly as possible by whatever means necessary. The computer may seem a little more magical and quite a bit more mystifying than other tools. But from the user's perspective, it is not essentially much different from a stapler, say, or a paper knife or card box.

The more complicated a machine is and the more miscellaneous functions it acquires, the wider the gulf between the engineer and user becomes. The basically very limited role the engineer plays in designing the machine makes it almost impossible for him or her to imagine all the various applications the user will want to find for it.

Oh, miserable users! Oh, miserable machines! The Japanese computer companies kept on sending their products out onto the market accompanied by meaningless manuals some seven inches thick, filled with countless pages of disorganized jargon. Users purchased the machines with bright visions of a shining future in which life was made not only easier but actually enjoyable by this modern, efficient equipment. A glance at the first few pages of the manual was all it took for these lovely visions to disappear. The users rushed to the phone, only to hear the busy signal at the customer-service center. Little wonder, then, that it is not at all unusual to find piles of discarded PCs in the junk yards of the major cities of this country.

The first to become aware of the need for better manuals were the software publishers. Software, after all, has little of the natural glory of hardware. Whereas some PCs make very attractive room ornaments even if they are never used, the value of a software program only becomes apparent when it is run on a computer. Software does not look the slightest bit impressive when it is lying asleep in a floppy disk jacket.

When experience had taught them that large numbers of complaint calls could absorb their engineers' time, some software publishers started to employ writers or to contract work out to writing services. Just as in the United

States, in Japan commercial book publishers also saw the opportunity and began adding PC software manuals to their catalogs.

In Japan today, hardware manufacturers are just beginning to follow this trend. Almost none of them had a dedicated documentation division until 1985 or 1986. And the rare exceptions were inevitably those who had jumped into the business from different fields such as home appliances or office machines. Faced with keen competition in the PC market and an increasing number of complaints from their domestic customers and dealers, the hardware manufacturers finally started installing in-house organizations for the sole purpose of producing documentation for users.

BEARING THE BURDENS OF AN EMERGING PROFESSION

Those of us who got in on the ground floor of the technical writing profession in Japan, who have gone to great lengths to encourage its growth and shape its development, cannot help but feel a certain amount of satisfaction as we witness its emergence today as a full-fledged profession. But we still have a long way to go before that satisfaction can be anything but tentative. Writers here still have a much too heavy burden to bear.

Five years ago virtually no one in Japan even knew what a *tekunikalu laitah* (technical writer) was. Today the situation is much improved, with major Tokyo newspapers carrying perhaps one classified ad or so a month announcing openings in the field. Nevertheless, when you answer an ad and get one of these jobs, you find you have nothing to rely on except yourself. You are probably working under a supervisor who has never written a single page of a manual. The style guide you are given, if any, has not been revised since it was put together some twenty years ago. And there are no textbooks available, not even at the largest bookstores.

Perhaps a few night classes would be in order? Yes, if there were any to be had. But there is absolutely no university or college in the entire country that offers a technical writing class. Consequently, there are no academic researchers in this field, either. It is only within the last year that a few private, once-a-week seminars have been offered for prospective writers. At them, lectures or speeches are given by active technical writers with widely varied skill levels and experience. Better than nothing, perhaps, but this hardly counts as education.

To make matters worse, there is no professional organization to turn to, either. For practicing writers, meeting with other professionals in the same field can often afford a better learning opportunity than participating in seminars. A few of us—writers, editors, and others—who share an interest in documentation have begun to meet regularly to discuss various topics related to our field. But this is still a very limited forum for exchanging views and experiences. We are looking forward to the establishment of a professional organization such as the Society for Technical Communication in Japan.

Another problem technical writers face here involves ridiculously tight schedules. Top management typically believes documentation production costs should come out of the hardware component budget rather than granting documentation an independent budget of its own. As a result, the writer's efforts in planning, defining the audience, and deciding the contents and organization are invariably underrated. "You have three weeks to finish this 200-page manual," your manager says. And fighting a manager like that could easily ruin your nerves.

One of the saddest things about the technical writer's situation in Japan is that he or she seldom receives respectful treatment. A former programmer I know at one large company was promoted recently (at least, the company said it was a promotion) to the position of technical writing supervisor. But when he showed his new business card to his wife, her response was not all he could have desired. "What have you done wrong in the programming division?" she shrieked.

All of this leaves us with just a single—albeit far from insignificant—source of comfort and hope: the user can tell the difference.

LEARNING FROM THE WEST—AND ADAPTING WHAT WE HAVE LEARNED

Lacking models at home, we turned to the west, especially to America, and found we could learn a great deal from professionals across the Pacific. At the risk of oversimplifying, I would suggest that there have been three areas in which what we have learned has been of particular significance.

Tutorial and reference. The first thing we learned, and one of the most significant, was the importance of altering our explanation to suit the level of the user. We also learned that the user's level should be measured in a variety of ways, taking into account not only knowledge and experience but also such things as interest in the product and the benefits expected from it.

Overview. Our second lesson involved the importance of starting every book, chapter, section, and paragraph with an overview. We learned, in other words, to show the reader what we were going to explain before getting into detailed descriptions of procedures.

Machinespeak. The third very helpful lesson we gained from our observation of Western methods is that it is always best to let the machine talk whenever possible instead of creating a book. This means both creating online documentation and making suggestions to the research and development division in order to improve the product itself.

It was not long, however, before we realized that even the best Western models were not totally suited to the Japanese user. Japanese may sometimes make too much of their differences from Westerners, but it is nevertheless clear that a number of differences do exist.

Learning process. One of the chief differences we discovered involves the way we Japanese approach learning. We do not want a verbal overview; we want it to be presented visually. A typical Japanese user response might be, "Don't go to great lengths to tell me how wonderful it is. Just show me what is so great about it."

Reading habits. Despite the pride so obnoxiously expressed by a recent prime minister about the fact that Japan's illiteracy rate is the lowest in the world, the Japanese, in general, hate to read any kind of documentation. A similar tendency probably exists in the United States as well. But here in Japan the case is extreme. Words like "manual," "contract," or "regulations" are no more than alternative ways of saying "dull" or "a waste of the time you spend trying to understand it." This means that manual writers in Japan have to put the greatest possible effort into making sure their products never look like manuals.

Keyboard usage. Written Japanese includes two different sets of phonetic symbols, each consisting of fifty basic characters (some with uppercase and lowercase characters, some without), and one type of ideographic symbols (Chinese characters or *kanji*) consisting of some seven thousand characters, in addition to the Western alphabet. Since a variety of keyboards have been developed to generate these characters on word processors or computers, we cannot expect our reader to understand even the basic keyboard layout, as the Western writer can. When one writes for a completely novice audience, especially, keyboard explanation inevitably becomes a fat section in Japanese software manuals.

Sales and distribution. A final important difference between the Japanese computer user and his Western counterpart involves sales and distribution procedures. It is said that the average distribution cost for industrial products in Japan is roughly twice that in the United States. Japanese customers expect more individual service from sales or customer-service personnel. And these personnel are, in fact, far more responsive than those in the United States, for example. Accustomed to special treatment and prone to assume they can get all the information they need orally from the manufacturer, Japanese make especially reluctant readers. They have to be entertained or in some other way lured into a manual—and then teased into staying there.

All of these factors required that we search for new ways to approach our Japanese readers. Western manuals had carried us a long way, but we would have to go the rest of the way ourselves. In my company we began to experiment, taking great care to document and share everything we learned. The manuals we create now differ from Western manuals in a few major ways.

To begin with, we always show the features of the product first. Rather than starting with a written overview such as is commonly found in Western manuals, we give the user a visual overview. Thus we may provide at the beginning of the manual a typical example of the product's output. If the product is a portable word processor, for instance, we put a printout sample employing the word processor's features on the very first page of the manual. This acts as a showcase, enabling us to reveal the machine's benefits directly, rather than explaining how the user should apply it to get the benefits he wants.

We also like to use the manual to make users comfortable with the product and create a positive attitude toward it by appealing to their likes and dislikes. Japanese like cartoons, so we give them plenty. Fashion magazines sell well here, so, yes, we sometimes use illustrations in the fashion magazine style. They hate manuals? Okay. We put a cover on ours that does not look anything like any manual cover they have ever seen to make sure they are not reminded of bad experiences in the past.

The fact that we are dealing with a very complicated written language—so complicated that only experts could type it until the word processor came along—also makes a big difference. I mentioned above that we have to devote a great deal of space to explaining the keyboard. For the same reasons, we start the tutorial with a keyboard trial, being especially sensitive to the needs of novices.

The nature of our language, in which the word order is often different from that of English, also means that it makes no sense to arrange the reference material in alphabetical or phonetic order. As a result, we always group the commands and functions by user benefit.

One conclusion that can be drawn from this, of course, is that foreign companies wishing to introduce their products to Japan should avoid the direct translation approach. Their manuals, not to mention their screen commands and help messages, should be completely rewritten from scratch rather than translated. Cultural and socio-political differences require not only a complete reorganization of the original manual, but also essential changes in the approach used as well as the insertion of additional information.

I would like to think that the day will come when some of the things we are doing here will contribute to the improvement of writing methods in the West. For this to occur, however, we will have to create many more opportunities for international exchanges of views and experiences among technical communication professionals.

REFERENCES

Papers and monographs written in Japanese are not included in this reference list.

Bardre, Albert and Ben Shneiderman, eds., *Directions in Human/Computer Interaction,* Ablex Publishing Corporation, Norwood, NJ, 1982.

Green, Thomas R. G., Stephen J. Payne, and Gerrit C. Van der Veer, eds., *The Psychology of Computer Use,* Academic Press, New York/London, 1983.

McGehee, Brad M., *The Complete Guide to Writing Software User Manuals,* Writers Digest Books, Cincinnati, OH, 1984.

Norman, Donald A., Stephen W. Draper, eds., *User Centered System Design,* Lawrence Erlbaum Associates, Hillsdale, NJ, 1986.

Price, Jonathan, *How to Write a Computer Manual: A Handbook of Software Documentation,* Benjamin Cummings, Menlo Park, CA, 1984.

2

THE LANGUAGE OF PLAIN ENGLISH

Plain English: The Remaining Problems

Joseph M. Williams
The University of Chicago

Those who must write clear, readable sentences now have access to about 90% of everything they need to know to write those sentences. That knowledge can be encapsulated in three principles: (1) Express crucial actions as verbs; (2) locate the participants of those actions in the subjects of the verbs; (3) arrange information in those sentences so that older, more familiar precedes newer, less familiar. But if some scholarly knowledge is available, other needed information is not. We know relatively little about the nature of form. More importantly, we know relatively little about how best to teach the knowledge we do have. And most importantly, we know less yet about how to train those responsible for managing the prose of others. There is, though, one crucial piece of certain knowledge that takes precedent over all the rest: before we can train those in an organization how to write plain English, the highest levels of management must make it clear by their active participation in the program that clear communication is their highest priority.

Not too long ago, this item appeared in the *New York Times* News of the Week in Review, at the end of a report on litigation over GM's X-cars and their brake problems:

> In a recent report, the General Accounting Office . . . found that public recall notices are difficult to understand, which may contribute to the poor consumer participation — about 50 percent — in recall programs. (March 6, 1983)

Given our level of national literacy, the GAO's concern is probably not misplaced. I suspect they had in mind recall letters such as this one, which I in fact received:

> A defect which involves the possible failure of a frame support plate may exist on your vehicle. This plate (front suspension pivot bar support plate) connects a portion of the front suspension to the vehicle frame, and its failure could affect vehicle directional control, particularly during heavy brake application. In addition, your vehicle may require adjustment service to the hood secondary catch system. The secondary catch may be misaligned so that the hood may not be adequately restrained to prevent hood fly-up in the event the primary latch is inadvertently left unengaged. Sudden hood fly-up beyond the secondary catch while driving could impair driver

Reprinted by permission of the publisher, *Visible Language,* Rhode Island School of Design, Providence, RI 02903.

visibility. In certain circumstances, occurrence of either of the above conditions could result in vehicle crash without prior warning.

While I found alarming the news of my possible "crash without prior warning," the quality of the text was only a little less so, since I could imagine someone less than entirely literate trying to get through it, discarding it, and increasing, however slightly, *my* chances of confronting that person's episode of acute hood fly-up.

Now there is no mystery why this letter is so badly written. The remote causes have to do with the corporation's twin fears over its customers' anxiety and its own liability: the manufacturer wanted to cover itself against lawsuits, but not enough to risk alarming possible repeat buyers. The writers also probably had to deal with language mandated by the government and with the difficulty of writing documents by committee. To speculate whether its motives might also have included minimizing the costs of a massive recall is perhaps unnecessarily cynical.

The proximate cause of this obscurity, however, is also perfectly clear: The writers systematically, perhaps deliberately, constructed a document that violated three fundamental principles of clear sentences. In almost every instance, the writers expressed critical action not in verbs but in abstract nominalizations, deleted agency, and buried the most significant information in the middle of sentences. As a result, the text became abstract, impersonal, and cognitively demanding. Here is what I mean. On the left are the verbs in the passage; in the middle are the actions, on the right are all the actions expressed as verbs:

	Verbs	*Actions as nouns*	*Actions as verbs*
(1)	involve		
(1)		failure	fail
(2)	may exist		
(3)	connects		connect
(4)		failure	fail
(4)	could affect		
(4)		directional control	not steer
(5)		brake application	brake
(6)	may require		
(6)		adjustment service	adjust
(7)	may be misaligned		not align
(8)	may not be restrained		not restrain
(8)	to prevent		not prevent
(8)		fly-up	fly up
(9)	is left unengaged		not engage
(9)		fly-up	fly up
(10)		driving	drive
(10)	could impair		
(11)		visibility	not see
(11)		occurrence	occur

(12)	could result		
(12)		crash	crash
(12)		warning	not warn

In almost every case, the writers expressed the potentially damaging part of their story in nominalizations and passives, and with three exceptions (*your vehicle,* twice; *driver visibility),* deleted all references to the cast of characters. If we merely undo those three operations — if we express crucial actions as verbs, make the verbs active, and restore the cast of characters in their subjects — the passage becomes not just more readable, but appropriately more threatening:

> Your car may have a defective part. If this part fails, you will not be able to steer, especially if you brake hard. We may also have misaligned the secondary catch on your hood. If you don't engage the primary latch, your hood could fly up. If the part fails or your hood flies up, you could crash.

To edit in this way requires a very slight level of skill and judgment. And having said that, you will understand why I was not pleased to read the sentence that concluded the *Times* article: "The agency has given a linguist five months and $23,000 to write an understandable form letter." Somewhere in the bureaucratic bowels of the Federal government there stirred an impulse toward a scientific, empirical, *data*-based argument that letters such as this one are less than entirely reader-friendly. It is an impulse by no means foolish, or even contemptible. But it is a great waste of the taxpayer's money. And it is a striking bit of evidence of how incompetent some are in their inability to read a text, recognize that it is marginally comprehensible, and to say so without needless statistical support. We do not need five *minutes* to rewrite this letter, much less $23,000. We do not need five months of testing to recognize that by applying almost mechanically a few simple principles of style, we have produced something entirely comprehensible.

In fact, I would like to make a claim that some of my more research-minded colleagues may find outrageous: At the level of sentence structure, we know about 90% of all we need to know to write plain English. And it does not include formulae by Flesch or Gunning or Dale-Chall or any of the legion of others who have reduced readability to numbers. For all *practical* purposes, we need only three simple principles of prose discourse:

(1) Express *crucial* actions as verb.
(2) Name in the subject of the verb one of the cast of characters involved in the action, the *agent* of the action, if possible.
(3) Put at the beginning of your sentence that information which links the sentence to what has gone before or which is relatively more familiar to your audience than anything else in the sentence; put at the end that information which is relatively more complex, less familiar, less predictable.

The first principle invites the second, because once you have arranged your information around specific verbs, you are more likely to make the subject of the verb one of the participants — usually the agent — in the action. And this second principle reflects the third, because the cast of characters is the most familiar component in any story. If they are put into the subject of the sentence, the newer information will by default fall out closer to the end.

I will simply assert — flatly — that any discourse that follows these three principles will be significantly more readable than discourse that does not. Such discourse may have other problems of style: unfamiliar vocabulary, interruptions, excessively multiplied clauses, confusing negatives. But I will simply assert again that these problems are less likely to appear in prose that meets the three criteria. Readability tests that measure word length, sentence length, etc., merely record the *symptoms* of overly complex prose. The *etiology* lies in the systematic dislocation of action away from verbs, agents away from subjects, and familiar information away from the beginning of the sentence.

Those who would demand numbers to verify their response are like a person lying in bed, shivering, vomiting, wracked by spasms of the gut who, when asked if he were sick, answered, "I don't know, I haven't taken my temperature yet."

In schematic terms, the principles can be collapsed into a hierarchical array:

topic	comment	
old	new	
subject	verb	object
agent	action	—

Figure 1.

Express familiar information in the Topic of the sentence (usually, but not always the subject). In these sentences, *style* is the topic, but in neither case the subject: *As for style, there is much to say. Style I know very little about.* Express newer, less familiar, more complex information in the Comment of the sentence, in that part of the sentence that follows the Topic. Use the Subject to name the Agent (or at least one of the cast of characters), and use the verb to express the crucial action involving that character. The rest of the sentence will for the most part take care of itself.

Now it is true that there are a good many other things about the nature of discourse about which we know very little. We do not know very much about how writers communicate their intentions. We know relatively little about how pre-existing knowledge influences how we read. We know relatively little about how the *reader's* intention in reading influences how he or she reads. We do not know the relative importance and interplay of the different components of discourse that contribute to a sense of form. In fact, we don't really know how we get from one sentence to the next, how we fill in what is not said. In fact, we are rather in the dark about the whole concept of coherence in discourse. But even if we do not know as much as we would like about these matters, we know enough to create documents more readable and usable than that automobile recall letter. And we should know enough not to have to test the revised one.

Yet even if we do know 90% of what we need to create readable, usable documents, we are, I think, ignorant of about 90% of what we need to know to teach, to communicate, to get people to learn and use this knowledge. For example, what is it that makes some among us genuinely unable to tell the difference between these sentences:

> This agreement may be terminated by either party without cause upon thirty days written notice to the other party. Further provided that such termination would not be effective to terminate any Advisory Agreement or to terminate any commitments made by the Bank as agent for an investor with termination of obligations under such latter Agreements or commitments to be terminated as provided in the Advisory Agreement and the commitment.

> Either party may terminate this agreement without cause 30 days after it notifies the other party in writing. However, if either party wishes to terminate the Advisory Agreement or any commitments made by the Bank as agent for an investor, the party must observe the provisions they contain.

In every one of our writing programs, some participants seem intrinsically unable to distinguish between clear and unclear writing, or if they can, unable to fashion a sentence, even with help, that improves substantially on a terrible example.

There are more complex issues. We all need the technical vocabulary, tone of voice, that makes us sound like one another's professional peers. We often disparage it, but to what degree is it in fact a psychologically or rhetorically effective device? With young professionals not yet thoroughly socialized into a field, it serves a perfectly apparent psychological purpose. This was perhaps best expressed by a very new lawyer in a writing program we had just completed. He said, without the slightest trace of irony, "We've just spent three years learning to sound like lawyers, and now you want us to sound like ordinary people."

Professional dialects serve a rhetorical purpose, despite the more general abuse they receive in the popular press. But what are the limits? How do we best approach the problem of what is essentially bi-dialectalism? Indeed, the research on the range of professional languages, the range of professional argumentation, is just beginning to become a substantial field.

And *when* do we teach all of this? The stereotypical argument at every level in the process of education is that the teachers at the prior level did not do their job, did not teach their students what *we* (at whatever level "we" might be) think they should be taught. Thus teachers of freshman English deplore high school education, faculty in advanced courses deplore freshman English, graduate schools deplore undergraduate education, and professionals regularly deplore the quality of graduates joining their ranks. We might conclude that if those high school teachers had just done their job, this paper would be irrelevant.

In fact, there is an increasing body of evidence arguing for the proposition that one cannot learn complex activities like writing, problem solving, critical thinking, argumentation, etc., as generic skills. The skills of a good writer, problem solver, critical thinker, or teacher of rhetoric are so deeply enmeshed in a total understanding of the field itself, so tightly woven in and around sheer knowledge about a subject matter, that we cannot reasonably expect to graduate from almost any level students ready to behave in an entirely competent way at the next. The demands of simply learning a broad subject, of being socialized into the universe of discourse, of becoming part of that universe are so severe that we ought not be surprised when those graduates appear to be less than entirely competent. Thus a new doctor, a new lawyer, a new engineer will predictably suffer through a period while he or she is learning to sound like a doctor, lawyer, or engineer — or freshman at college, or student just starting to major in psychology, or new graduate student.

And this brings us to a yet more difficult problem. As I said, it is easy for any of us here to teach these issues. We all have a professional interest in writing and teaching writing. Many of us act as consultants to large organizations that perceive a need for our assistance. But however lucrative that perceived need may prove to be for us, we must acknowledge that ultimately those on the scene will have to assume the responsibility of passing on whatever we teach. In short, we are professionally obligated to make ourselves obsolete. We must be able to train people not merely to write well, but to train others how to write well, to train others how to manage the writing of those for whom they are responsible.

In this regard, the range of managerial styles we have to deal with is quite varied. It ranges from accepting a document from a subordinate, rewriting it to fit what the manager wants it to be, and sending it out without ever telling the subordinate what was done to the subordinate's document, much less why. Slightly more helpful are those who rewrite and at least send a copy of the rewrite to the original drafter, on the assumption perhaps that merely by privately comparing originals and revisions the drafter will eventually figure out what he should be doing. Only slightly more helpful are those who send the document back with advice like "tighten up," "be more organized," "clarify," etc. It may be useful advice, but it is about as helpful as a doctor telling that person lying in bed trembling and vomiting that he should "get better," "be healthier." Accurate advice, but not very useful.

More helpful are those who return a document, edited, with some specific comments about the reasons for the changes in that particular document: "I changed this sentence to make the FCC the subject because you said you were going to talk about new FCC policy in this section." It is not a great leap to, "Make whoever is most responsible for the actions you are describing the subject of your sentence." But it is a leap most managers are unable to make, because they simply have never had the occasion to study language, discourse, or to articulate the generalizations that apply not just to *this* text to make it clearer, but to texts in general.

But that is the kind of managerial style that subordinates find most useful: Unless they are explicity instructed in principles of action general enough to rise above trivial issues but powerful enough to apply to a wide range of potential behavior, they will learn only by trial and error, or not at all. These are the principles we began with: actions as verbs, subjects as agents, old before new. They are general, powerful, and specific.

They are also the principles that those who are responsible for the writing of others must control. They must know them not as tacit knowledge, but as active, articulatable knowledge. And they must be able to connect them to their gut response to a text that prompts such responses as "I keep losing the thread of the argument here," "I just can't get through this stuff," "I keep having to skip back to the previous sentence." They have to be able to locate quickly *in the text* the cause of their response, to analyze it, and to explain it.

In short, *we* as professional educators, consultants, and writing teachers may control the abstract knowledge that underlies discourse. Indeed, we must. But if we are substantially to serve social and economic objectives we have to be able to pass that abstract knowledge on in a way that makes it seem relevant, useful, nonthreatening, concrete, even interesting and challenging.

Whatever else we may think we know or don't know about language and discourse, we know precious little about demystifying it for those who perhaps remember from their English classes only that bewilderment and despair attendant on diagramming a sentence more than 10 words long. Language study has too often been a terrifying experience for too many of us. Replacing that deep dislike with an informed and easy knowledge of how discourse works is no small task.

We have been reassured that one way toward that end is through the use of computers. We are told that programs such as *Writer's Workbench* or *Epistle* will analyze our prose for us, point out our possible weaknesses, and even teach a lesson on correcting them. From what I have seen of these programs, they are simultaneously too powerful and not powerful enough. A simple command gives us much more information than any reasonable person probably wants to know. We get a statistical profile of our passives, nominalizations, occurrences of *be, there, it,* etc. A statistical profile of my automobile letter would have told us that it was probably suitable for a college level audience. But I would think that the most modest education would allow one to come to that conclusion before it came up on the screen. We do not wait until we read a thermometer before we take an aspirin.

Certainly, such text editing programs may be useful for those who think they need a thermometer before they can tell whether they feel sick. There is an irony here: Such programs are attractive to those who are not skilled enough to work through a text quickly and efficiently. If they lack those skills, we must assume they lack the more sophisticated skills of understanding the powerful role of form on the way we process text. Given that more serious incompetence, how will the user use such programs? Surely not in the skilled way competent editors would. Given our experience of how such rubrics have been used elsewhere in our lives, we should not be surprised to see managers using these programs in ways that rely essentially on getting the right numbers out of the statistical profile.

What I want to conclude with has been in my experience the single most crucial aspect of any successful program in improving the discourse in an organization. We can control and pass on all the knowledge, all the skills, all the theory any manager might need to pass it all on in turn to his or her subordinates. We can teach managers how to handle the writing of others, how to analyze, edit, or explain. But unless we have been successful in achieving a prior objective, none of this will prove very useful, or enduring — at least it has not in the experience of my colleagues and me. That prior objective is obtaining from the most senior levels of management not just approval for a concerted effort to improve the quality of discourse in the organization, but a commitment that communicates itself through every component of the organization. Every senior manager wishes that those under him or her communicated better. But not every senior manager is willing to commit the time and effort to a program in communication to achieve that end. In short, in my experience, it has not been the technology, the knowledge, the skills, or the teaching situation that has made the difference between successful and unsuccessful programs in teaching written communication. What has made the biggest difference has been from senior management a commitment of time, of visible presence, of articulate and continued support *over a sustained period of time.*

My colleagues and I are a fairly successful group of consultants. Clients invite us back. But I am increasingly convinced that it is not because of what we know or how we teach it. It's because we repeat and repeat again how important it is that every manager of the group we are working with makes it clear to the group, *by participating in the group,* that what we are doing is not just important, but must and will be done. In fact, if truth be told, if the managers simply articulated their determination to improve the quality of writing in their organization, consultants would probably be out of a job.

This, then, is the remaining problem in the plain English movement: not more knowledge about the structure of discourse (though that is very important); not more knowledge about how to teach it (though that is more important yet); not more technology in the form of ever-more detailed text-editing programs (though that is important, too). The remaining problem is convincing those with the authority that teaching clear written communication does not begin with those who do the communicating. It begins with those who bear the ultimate responsibility for it; it begins with those with the authority.

Prescriptive Linguistics and Plain English: The Case of "Whiz-deletions"

Thomas N. Huckin, Elizabeth H. Curtin, and Debra Graham
Carnegie-Mellon University, Pittsburgh

The plain English movement is under considerable pressure to produce simple, easy-to-apply guidelines for writers. Often, however, such guidelines are not consistent with the actual practices of good writers and are therefore ineffective. As a case in point, this paper discusses the guideline "Avoid whiz-deletions," taken from a highly-acclaimed plain English handbook. It is shown that whiz-deletions, or reduced relative clauses, actually abound in good writing. They outnumber full relative clauses by a 4-to-1 margin in good standard prose and by an 8-to-1 margin in model plain English documents. Whiz-deletions are useful in helping to de-emphasize information, promote sentence rhythm, facilitate parsing, avoid ambiguity, and omit needless words. The maxim in its original form should be replaced by a more descriptively-accurate one that reflects these features. In general, the plain English movement should promote only those guidelines that are consistent with the practices of good writers. Furthermore, such guidelines should be formulated so as to direct the attention of novice writers to broader contextual factors, not just sentence-based ones.

Prescriptive linguistics is the study of how language is used by good writers, its purpose being to devise rules-of-thumb which can help other writers. In this sense, although its methods of analysis are borrowed from descriptive linguistics, sociolinguistics, psycholinguistics, rhetoric, and other primarily theoretical fields, its orientation is basically pedagogical. The plain English movement, devoted to improving the readability of public documents, also has an important pedagogical function. In trying to persuade lawyers, bureaucrats, businessmen, and others to produce writing that is easily understood by the general public, it is under great pressure to provide rules-of-thumb that are both easy to apply and effective in making writing more comprehensible.

Unfortunately, research in prescriptive linguistics (and its related field, readability research) shows that there are very few rules-of-thumb that satisfy both of these criteria. In their analysis of noun compounding, for example, Kaufer and Steinberg (1984) show that a simple maxim like "Avoid noun compounds," though easy to apply, is so oversimplified that it does not reliably improve readability. On the other hand, their revised maxim, though effective if faithfully followed, is significantly more difficult to apply. This trade-off relationship can be seen in other recent studies as well, such as Selzer (1983),

Reprinted by permission of the publisher, *Visible Language,* Rhode Island School of Design, Providence, RI 02903.

Huckin (1983), Moskovit (1983), Geisler, Kaufer, and Steinberg (1984), and Steinberg (1986).

In our view, there is no obvious way around this difficulty. It is important for the plain English movement, however, not to succumb to demands to produce oversimplified, ineffective prescriptions. No matter how easy a maxim may be to apply, if it does not result in better comprehension then it can serve no useful purpose. And what is worse, the hard-earned credibility of the movement will suffer. It is better, we feel, to produce accurate descriptions of what good, clear writing actually looks like. Though such descriptions may turn out to be quite complex, at least we will have a more accurate picture of what plain English is and will be able to go about devising appropriate methods for teaching it.

Current practice in plain English pedagogy.

To illustrate our contention that stand-alone prescriptions tend to be oversimplified and ineffective, we will focus our attention on a prescription drawn from *Guidelines for Document Designers* (henceforth *GDD),* the state-of-the-art handbook published by the American Institutes for Research. Written by five well-known experts in the field (Daniel Felker, Frances Pickering, Veda Charrow, Melissa Holland, and Janice Redish), *GDD* contains what we feel is the most complete and most representative set of plain English principles yet produced. As the authors state:

> We chose the document design principles in these guidelines for two reasons. First, experienced document designers widely agree that the principles in this book do influence the clarity of prose. Indeed, all of the principles we have selected are included in at least several writing and design manuals aimed at practitioners. Second, all of the principles in these guidelines have been examined in research studies to see how they affect the comprehension of information. (p. 1)

Based on research that is summarized in an earlier book by this same team *(Document Design: A Review of the Relevant Research,* 1980), *GDD* contains twenty-five maxims for writers of public documents. These maxims include principles for organizing text, for writing sentences, for formatting, and for using graphics. For each maxim, the research on which it is based is explicitly described. It is this feature that makes *GDD* stand out from other writing handbooks. While the maxims themselves are quite similar to those found in Strunk and White's *Elements of Style,* the *Harbrace College Handbook,* and many other writing handbooks, the citing of specific psycholinguistic and cognitive psychological studies makes *GDD's* treatment unique.

One of *GDD's* maxims is "Avoid whiz-deletions," which it describes as follows:

> In English, many subordinate clauses are introduced by the words "which is," "who were," "that are," etc. These "little words" help make the structure of the sentence clear to the reader — they make it easier for the reader to understand how the subordinate clause relates to the rest of the sentence.

> Removing these "little words" is called "whiz-deletion." "Whiz-deletion" can often make a sentence unclear or ambiguous; it can place a greater than necessary burden on the reader. Wherever possible, replace the missing "which is," "who was," etc. [*GDD,* p. 39]

GDD exemplifies this statement with these two sentences:

(1) a. The director wants the report *which was* written by the Home Office.
   b. The director wants the report written by the Home Office.

In this example, deleting the relative pronoun and auxiliary verb produces — out of context — an ambiguous sentence. Instead of the meaning given in version 1a, it could mean "The director wants the report *to be* written by the Home Office." With this in mind, the authors of *GDD* say, "To avoid ambiguity and to reduce the burden on your readers, restore the missing relative pronoun plus copula verb *whenever possible*" [emphasis ours]. This statement amounts to a strong prescription: "never leave out a *which is, that are,* etc." The authors of *GDD* later qualify this prescription somewhat when they write: "It is not necessary to *always* restore the relative pronouns plus copula verb. If the whiz-deletion does not interfere with the clarity of the sentence and the reader's ease in reading it, you may leave the sentence as it is." But they clearly imply that such instances of "harmless" whiz-deletion are the exception, not the norm.

Noticing that the ambiguity of sentence (1b) results not from whiz-deletion alone but also from the presence of a main clause desiderative verb *(wants)* which can take either a sentential complement or a noun-phrase complement, we were curious to know how common such sentences actually are in real-world writing. Do they occur frequently enough to merit a maxim? It is possible that whiz-deletion leads to ambiguity in other kinds of sentences as well, but the other two example-sentences that *GDD* provides are not compelling ones.[1] The authors of *GDD* say only that "whiz-deletion can often make a sentence unclear or ambiguous"; they do not cite any evidence.

GDD does refer to psycholinguistic research, claiming that it gives "indirect" support for the maxim. But when we looked closely at this research, we found that it had very little relevance for whiz-deletion. For example, Fodor and Garrett (1967) and Foss and Lynch (1967) used sentences with double embeddings, as in "The pen (which) the author (whom) the editor liked used was new." Such sentences almost never occur in real life and, furthermore, do not contain whiz-constructions. Hakes and Cairns (1970) used sentences with only single embeddings, but these sentences did not contain whiz-constructions either. Hakes (1972) used sentences which had no relative pronouns at all, only complementizers (e.g., "The blind student felt (that) the material in the art course would be too difficult for him to understand."). Charrow and Charrow (1978) did study whiz-deletions, but — like all of the studies cited above with the exception of Fodor and Garrett — their experimental procedure used oral rather than written texts and thus did not measure readability.

Survey of data.

Given the fact that there is virtually no psycholinguistic evidence for or against the use of whiz-deletions, we thought it would be useful to consider a different kind of evidence — namely, the practices of good writers. Do good writers use whiz-deletions? And if so, why? In answering these questions, we first looked at samples of plain English, then we looked at samples of unsimplified writing.

Our samples of plain English were taken from *How Plain English Works for Business: Twelve Case Studies,* published by the U.S. Department of Commerce. We chose this collection because it is recently published (1984), because it contains writing samples that are considered models of plain English, and because it was edited by the same people who wrote *GDD* and devised the "Avoid whiz-deletions" maxim. For every sample of plain English in this book, we counted the total number of words, the number of whiz-deletions, and the number of "whiz-constructions" (i.e., full whiz-forms without deletion).

The first document we examined from this volume was a consumer promissory note from Citibank, approximately 550 words long. It contains 7 whiz-deletions and one whiz-construction. Some examples of each type are (▲ marks the deletion side and boldface marks the whiz construction):

(2) I understand that I must maintain property insurance on the property ▲ covered by the Security Agreement for its full insurable value.

(3) You can then demand immediate payment of the balance of this note, minus the part of the finance charge **which hasn't been** earned ▲ figured by the rule of 78.

The second document is a consumer information guide from JCPenney. It is approximately 770 words long, contains 3 whiz-deletions and one whiz-construction. Here are examples of each type:

(4) If you are fairly active, under age 35, and healthy, you can develop your own exercise program from guidelines ▲ given in this guide.

(5) Exercise cycles are one-wheeled stationary bicycles **that are** pedaled like a regular outdoor rolling bicycle.

There are 8 other samples of plain English in this collection of case studies. They, and the documents from Citibank and JCPenny, are listed in Table I along with the relevant data-counts. It is clear from these data that the guideline "Avoid whiz-deletions" is not observed in model samples of plain English.

When we looked at samples of writing from *GDD* itself, we found that it does not adhere to its own guideline. A random selection of 9 pages, totalling about 3000 words, revealed 11 whiz-deletions and only 2 whiz-constructions.

Broadening our survey, we looked at randomly-chosen samples of good but unsimplified writing from a variety of published sources. We examined technical writing (a memo, a business letter, two process reports, and an investigative report) from Kolin and Kolin's *Models for Technical Writing,* scholarly journals (e.g., *The New England Journal of Medicine, College Composition and Communication, Child Development, Memory and Cognition, Current History),* advertisements for IBM, AT&T, Apple, Planned Parenthood, Sebastian International, and the Collector's Library (found in *Time, Esquire, Atlantic Monthly, Glamour,* and *Popular Computing),* popular magazines *(National Geographic, New Yorker, Psychology Today, Science 84, Sports Illustrated),* informational writing *(Encyclopedia Britannica,* a plain English brochure), journalism *(Washington Post, Wall Street Journal, Time),* and fiction from Isaac Asimov, Philip Roth, James Michener, Joyce Carol Oates, and John Updike. The data is summarized in Table II.

We can conclude from all this that good writing in general, whether it is labeled plain English or not, does not follow the prescription to "Avoid whiz-deletions." In fact, whiz-deletions are so heavily used that they can be considered, if anything, *a standard feature* of good writing, with whiz-constructions

Document type	*total words*	*whiz-deletions*	*whiz-constructions*
St. Paul insurance policy	2760	35	6
Shell information sheet	1000	7	0
JCPenney information guide	770	3	1
Citibank consumer loan note	550	7	1
Pfizer healthcare ad	520	0	0
Shell ad	450	2	0
HOW homeowner's warranty	380	4	0
Roche's medication information	330	1	0
Insurance information brochure	110	0	0
Product safety notice	100	2	0
Aetna table of contents	100	1	0
	7070	62	8

Table I. Data for model plain English documents.

Type of writing	*total words*	*whiz-deletions*	*whiz-constructions*
Fiction	18185	41	6
Popular magazines	18156	68	21
Scholarly/academic	18038	121	18
Journalism	15209	66	18
Technical	2345	6	9
Advertisements	2298	16	4
Informational	1612	7	0
	75843	325	76

Table II. Data from good, unsimplified writing.

being used only for special circumstances. The reasons for this widespread disregard for the maxim are discussed in the next section.

Analysis of a sample.

When we analyzed our corpus for individual instances of whiz-deletions and whiz-constructions, trying to understand why the former are so preferred, we found some interesting patterns. First of all — and contrary to the claim made in *GDD* — we found that whiz-deletion almost never made a sentence unclear or ambiguous. There was only one case of desiderative-verb ambiguity like that in (1b),[2] and only one or two cases of ambiguity of other types. In fact, it was more often the case that whiz-deletion actually *reduced* ambiguity. Secondly, we found that whiz-deletion serves the major rhetorical purpose of de-emphasizing one part of the sentence so that another part can be emphasized. At the same time it often promotes better sentence rhythm. It can help make complex sentences easier to parse and thus easier to understand. Finally, it makes writing more concise — certainly a desirable feature for any plain English document.

These points are illustrated below with examples taken from a consumer loan agreement in *How Plain English Works for Business* (see Figure 1). In this pioneering case Citibank decided to devise a simplified promissory note that would be understandable to ordinary borrowers, not just to corporate clients. There are 7 whiz-deletions and one whiz-construction in this form. We will look at each in the order in which they appear. (The underscoring in all cases is ours.)

• *"Amount ▲Financed"* (line 12).

This phrase is part of a formatted list of items. There are many reasons why a writer might prefer the whiz-deleted form rather than a longer form such as "Amount that is Financed" or "Amount that is being Financed." First of all, clarity: the meaning of the expression is perfectly clear from the context, especially given the "1 + 2 + 3" formula that follows it in parentheses. Secondly, ease of reference: as a compressed form, it gives the word "financed" more of an adjectival (descriptive) quality than it has in the unreduced form, where it is more of verb. Thus it works better as a name or even as a technical term, and can easily be used as such by a loan officer or customer. Third, economy of expression: it uses fewer words and thus fits better into the space provided. Finally, parallelism: as a phrasal construction rather than a clausal one, it is similar in form to the other items on the list.

• *"you will refund the unearned finance charge, ▲figured by the rule of 78"* (line 18).

A consumer reading this loan note is not likely to be very concerned with the exact method by which bankers figure out an unearned finance charge. Hence, although this information may need to be explicitly stated for legal reasons, it will be treated by most consumers as a mere formality. The important point, for them, is contained in the main clause — that the bank will refund the

Consumer Loan Note Date________________________, 19____

(In this note, the words **I, me, mine** and **my** mean each and all of those who signed it. The words **you, your** and **yours** mean First National City Bank.)

Terms of Repayment To repay my loan, I promise to pay you ______________________________ Dollars ($__________). I'll pay this sum at one of your branches in __________ uninterrupted ______________ installments of $__________ each. Payments will be due ____________________ __________, starting from the date the loan is made.

Here's the breakdown of my payments:

1.	Amount of the Loan	$__________	
2.	Property Insurance Premium	$__________	
3.	Filing Fee for Security Interest	$__________	
4.	Amount Financed (1+2+3)		$__________
5.	**Finance Charge**		$__________
6.	Total of Payments (4+5)		$__________

Annual Percentage Rate______%

Prepayment of Whole Note Even though I needn't pay more than the fixed installments, I have the right to prepay the whole outstanding amount of this note at any time. If I do, or if this loan is refinanced—that is, replaced by a new note—you will refund the unearned **finance charge,** figured by the rule of 78—a commonly used formula for figuring rebates on installment loans. However, you can charge a minimum **finance charge** of $10.

Late Charge If I fall more than 10 days behind in paying an installment, I promise to pay a late charge of 5% of the overdue installment, but no more than $5. However, the sum total of late charges on all installments can't be more than 2% of the total of payments or $25, whichever is less.

Security To protect you if I default on this or any other debt to you, I give you what is known as a security interest in my ○ Motor Vehicle and/or ____________________ (see the Security Agreement I have given you for a full description of this property), ○ Stocks, ○ Bonds, ○ Savings Account (more fully described in the receipt you gave me today) **and** any account or other property of mine coming into your possession.

Insurance I understand I must maintain property insurance on the property covered by the Security Agreement for its full insurable value, but I can buy this insurance through a person of my own choosing.

Default I'll be in default:

1. If I don't pay an installment on time; or
2. If any other creditor tries by legal process to take any money of mine in your possession.

You can then demand immediate payment of the balance of this note, minus the part of the **finance charge** which hasn't been earned figured by the rule of 78. You will also have other legal rights, for instance, the right to repossess, sell and apply security to the payments under this note and any other debts I may then owe you.

Irregular Payments You can accept late payments or partial payments, even though marked "payment in full", without losing any of your rights under this note.

Delay in Enforcement You can delay enforcing any of your rights under this note without losing them.

Collection Costs If I'm in default under this note and you demand full payment, I agree to pay you interest on the unpaid balance at the rate of 1% per month, after an allowance for the unearned **finance charge.** If you have to sue me, I also agree to pay your attorney's fees equal to 15% of the amount due, and court costs. But if I defend and the court decides I am right, I understand that you will pay my reasonable attorney's fees and the court costs.

Comakers If I'm signing this note as a comaker, I agree to be equally responsible with the borrower. You don't have to notify me that this note hasn't been paid. You can change the terms of payment and release any security without notifying or releasing me from responsibility on this note.

Copy Received The borrower acknowledges receipt of a completely filled-in copy of this note.

	Signatures	Addresses
Borrower:	____________________	____________________
Comaker:	____________________	____________________
Comaker:	____________________	____________________
Comaker:	____________________	____________________

Figure 1. A revised consumer loan agreement from Citibank.

unearned finance charge. By omitting *which is* from the relative clause underlined, the writer reduces the predicative force of the clause and effectively de-emphasizes it. This in turn allows more emphasis to be given to the main clause, where it belongs. And since the whiz-words *(which will be)* are quite predictable from the context, deleting them reduces wordiness without causing any ambiguity.

Furthermore, when readers (including the writer reading his or her own writing) subvocalize, they would find that this sentence has better sentence rhythm without the whiz-words. The only words likely to be given intonational prominence in this phrase are *figured* (secondary) and *78* (primary). The latter gets end-focus[3] and also nuclear stress[4]; the former carries less informational weight but would probably receive a rhythmic stress due to its position. The word *rule* is not likely to receive any stress at all, since *78,* with which it is closely linked, gets nuclear stress. Thus there are eight unstressed syllables in this phrase. If the full whiz-construction were used, there would be ten or eleven unstressed syllables. With only two stressed syllables in the phrase, this would simply be too much of an imbalance.

- *"I give you what is known as a security interest in my* 0 *Motor Vehicle and/or __________________ (see the Security Agreement I have given you for a full description of this property),* ○ *Stocks,* ○ *Bonds,* ○ *Savings Account (*▲ *more fully described in the receipt you gave me today)* (lines 23-26).

Viewed in isolation, this might seem to be one of those rare cases where whiz-deletion creates ambiguity. The underlined relative clause, in principle, could be interpreted as modifying any of several possible referents: "Savings Account," "security interest," or "Motor Vehicle and/or . . . Stocks, Bonds, Savings Account." However, in actuality, there is very little possibility of such ambiguity arising. Immediately prior to reading this note, the consumer will have turned over to the bank the deeds to those properties that will be used by the bank as security, and received a receipt for them. Then, while reading the note, and just before encountering the relative clause, he or she will have put ticks in the circles referring to those properties. These properties, therefore, will be activated referents in the reader's mind, and will easily be associated with the relative clause. Hence there is no ambiguity.

But there will also be considerable variation from one consumer to another as to the number of circles checked. In some cases there may be only one; in others there may be two or more. If the writer wanted to use the full whiz-construction, he or she would thus have to use an awkward slash form of the auxiliary (". . .which is/are. . .") to cover both contingencies. By omitting the auxiliary verb altogether, i.e. by using whiz-deletion, the writer can gracefully avoid this problem. Furthermore, like the preceding case, this relative clause is of minor importance in this context and should be de-emphasized anyway. Reducing the length and syntactic complexity of the expression helps do that.

- *"I understand I must maintain property insurance on the property ▲ covered by the Security Agreement for its full insurable value"* (lines 27-28).

Here the situation is just the opposite of the preceding one. Using a full whiz-construction here, instead of reducing possible ambiguity, would actually increase it. The ambiguity arises from the role of the sentence-final prepositional phrase *(for its full insurable value)*. Does this phrase modify *must maintain property insurance* or does it modify *covered?* Psycholinguistic experiments show that readers who lack semantic knowledge and must rely more heavily on syntactic cues, i.e., nonspecialist readers, are likely to interpret the prepositional phrase (PP) as modifying the nearest predicate.[5] If *covered* were embedded in a full whiz-construction *(which is covered by . . .)*, it would be a full predicate and would cause many such readers to misinterpret the role of the modifier. Yet, according to our specialist informant, the PP is supposed to modify the earlier predicate, *maintain property insurance.* Whiz-deletion in this case serves to reduce the predicative force of the relative clause and thus make it more likely that the reader will correctly associate the PP with the earlier predicate.

Another way in which whiz-deletion promotes correct parsing in a case like this is by changing the phonological structure of the sentence, specifically, the location of likely breath-pauses. In the whiz-deleted version, the sentence is easily divided into three groups of roughly equal phonological length: *I únderstand I múst maintain próperty insurance / on the próperty covered by the Secúrity Agreement / for its fúll insúrable válue.* Not only is there a tripartite division here, but the middle breath-unit can easily be read with one less beat than the other two, as indicated, and is thus effectively subordinated. This makes it easier for the reader to see that the sentence-final PP and the main predicate go together. It also makes it easier for the reader to see that the main predicate and the final PP are the most important parts of this sentence. In effect the final PP is part of the predicate and shares its predicative force. Together they comprise the focal points of the sentence, whereas the intervening relative clause merely clarifies, for legal reasons, what property is being talked about. On the other hand if a full whiz-construction were used, the middle breath-unit would have more phonological weight than the other two breath-units (i.e., more syllables and just as many beats). This is contrary to the general principles of sentence-rhythm in English (Quirk et al: 1033-1052) which call for more weight to be given to end-units. Therefore many readers would try to rephrase the sentence as follows: *I understand I must maintain property insurance on the property // which is covered by the Security Agreement for its full insurable value.* This would be a nicely balanced rephrasing (19 syllables and 4 stresses followed by 21 syllables and 5 stresses). But it would induce many readers to interpret the final PP as attached to and thus modifying the relative clause, not the clause preceding it.

- *You can then demand immediate payment of the balance of this note, minus the part of the finance charge* <u>*which hasn't been earned*</u> *figured by the rule of* 78 (lines 33-34).

This is one instance in the text where a full whiz-construction has been used. The writer could have used whiz-deletion *(minus the part of the finance charge not earned figured by the rule of 78)* without much risk of ambiguity, or she could have used the same expression she used earlier in line 18 *(minus the unearned finance charge figured by the rule of 78)* and later in line 40 *(the unearned finance charge)*. But either of these forms would have taken emphasis away from the word *earned.* In the whiz-construction version this word receives full predication and phonological prominence (most readers would make a breath-pause right after this word, highlighting it and setting off the following phrase as a separate breath-unit; this will be discussed further below). This bestowing of emphasis could be accidental — merely a result of the writer attempting to use elegant variation. But elegant variation is not in keeping with the principles of plain English. Therefore, it seems more likely that the writer chose this form for a specific reason, that is, to deliberately give prominence to *earned.* In any event, whether done deliberately or not, the emphasis given to this word is certainly appropriate, for it has particular rhetorical value in this context. It occurs in the section defining default conditions and penalties. Since the bank is specifying its right to certain payments from the borrower, an antagonistic situation is depicted where a potential borrower (i.e., someone reading this note and deciding whether or not to sign it) would want some reassurance that the bank would not take money that it has not "earned." At the same time, by explicitly identifying part of the finance charge as having not been earned, the bank brings to mind the other part, that which has been earned. This is a subtle way of telling the potential customer that the bank will be working for him, not just taking his money.

- *You can then demand immediate payment of the balance of this note, minus the part of the finance charge which hasn't been earned* ▲ <u>*figured by the rule of 78*</u> (lines 33-34).

This phrase is identical to the earlier one in line 18. By now it can be seen as a formulaic expression, carrying a single holistic meaning. The principles of sentence intonation (Quirk et al.: 1033-1052) would dictate that *78* be given stress but not the other words in the phrase. This leaves nine unstressed syllables (versus the single stressed one). Using a whiz-construction would make the phrase even more unbalanced, with eleven unstressed syllables. Whiz-deletion, therefore, promotes sentence rhythm here even more than in the earlier cases.

Though not punctuated as such, this phrase is essentially a nonrestrictive relative, playing a parenthetical role in the sentence. If asked to read the sentence aloud, most readers, we believe, would pause briefly between *earned* and *figured.* (If an actual comma were inserted here, however, the phrase could be misinterpreted as modifying *the balance of this note.)* As in the earlier case, it is not the focal point of the sentence but just a formality, stipulating a method of calculation. By using whiz-deletion, the writer de-emphasizes it and allows more emphasis to be given to the phrase preceding it.

Another reason for using whiz-deletion here is to avoid confusion and possible ambiguity. If a full whiz-construction had been used, there would be two *which*- clauses in a row, with the referent of one *which* being embedded within the other. "You can then demand immediate payment of the balance of this note, minus the part of the finance charge *which hasn't been earned which is figured by the rule of 78."* The first *which* would refer to "the part of the finance charge," the second would refer to "the part of the finance charge which hasn't been earned." But many readers, we believe; would not be able to easily and quickly make this interpretation.

- *If you have to sue me, I also agree to pay your attorney's fees* ▲ *equal to 15% of the amount due, and court costs* (lines 40-41).

The writer could have used a full whiz-construction here ("I also agree to pay your attorney's fees *which will be equal to 15% of the amount due"),* but this would have highlighted the 15% figure and taken some emphasis away from the main clause. As in earlier examples where "the rule of 78" is a relatively minor detail, here the main point of the sentence is not the method of calculating the attorney's fees but rather the fact that it is the borrower who will have to pay them. Using whiz-deletion here not only helps downplay the relative clause but also saves words without any risk of ambiguity.

Rhythmic factors also play a role in favoring whiz-deletion here. This sentence occurs in a context (see Figure 1) where the focal information is carried by the words *have, sue, also, attorney, 15%, amount due,* and *court.* Of the other content words, *agree* and *pay* are "old" information repeated from the immediately preceding sentence; *fees* and *costs* are predictable after *attorney's* and *court* respectively (both are technical terms); and *equal* is virtually redundant from the context. Putting primary stresses on the focal words produces a nicely balanced rhythm (if read in context) without having to elide any unstressed vowels: *If you háve to súe me // I álso agree to pay your attórney's fees // equal to fifteén percént of the amòunt dúe // and coúrt costs.* The longest stretch of unstressed vowels in this sentence is . . . *fees // equal too* . . . Inserting whiz-words here would almost double the number of unstressed vowels in this sequence, and throw the sentence rhythm off. Some readers would perhaps try to compensate for this awkwardness by putting stress on *equal,* but this creates another kind of awkwardness, that of giving emphasis to a word that has little informational importance.

- *If you have to sue me, I also agree to pay your attorney's fees equal to 15% of the amount* ▲ *due, and court costs.* (lines 40-41).

This whiz-deletion occurs in the same construction as the one just discussed and was probably used for the same reasons. It de-emphasizes this part of the sentence in favor of the main clause, and it saves unnecessary words. Furthermore, *amount due* is essentially a technical term in this context (similar to *amount financed* in line 12) and is likely to be understood as a single lexical unit.

Discussion.

We have seen that whiz-deletions are used in this plain English document to de-emphasize information, to promote sentence rhythm, to facilitate parsing and avoid ambiguity, and to omit needless words. Whiz-deletion is used similarly in almost all of the 35 other well-written documents (plain English and non-plain English) we looked at. Clearly, it is seen by good writers as a very useful device, especially for de-emphasizing information and for creating graceful (and often useful) sentence rhythms. Furthermore, its use conforms to the standard handbook maxim of avoiding wordiness (even *GDD* has such a maxim: "Avoid unnecessary words"). Therefore, we must conclude that *GDD*'s prescription to avoid whiz-deletions is simply unjustified.

A more accurate approach, judging from the evidence that we have cited, would be to treat so-called whiz-deletion as the standard choice of good writers in most circumstances, with the full whiz-construction being reserved for special conditions. A revised maxim might then look like this:

"Use whiz-deletions except when:
(1) the word that follows the whiz needs emphasis, or
(2) the whiz-words help create better sentence rhythm, or
(3) the whiz-words are needed to avoid ambiguity (particularly following desiderative verbs)."

Our sample analysis has shown, however, that contextual factors (textual, rhetorical, other) play a major role in deciding what constitutes emphasis and what constitutes ambiguity. Anyone using this maxim, therefore, should look closely at each instance of usage and not just apply the maxim blindly.

Conclusion.

We wish to conclude this paper by raising — and trying to answer — the following question: Why did the authors of *GDD*, who are well-known experts in document design, make the mistake of devising the guideline they did? We think that it is the result, ironically, of the very strategy which sets their book apart from others — namely, a desire to ground their principles in rigorous, process-oriented empirical research. Certainly, such a move was badly needed. The problem is that it was simply not enough. The research they drew on in this particular section of the book comes almost exclusively from the sentence-oriented psycholinguistic tradition of the 1960s and 1970s. For the most part, such research examines only isolated sentences; it seldom considers the effect of context or of nonlinguistic factors on sentence comprehension. Since readers under normal circumstances rely heavily on context, on world knowledge, and on other broad-based linguistic and nonlinguistic variables in interpreting written documents, guidelines for writers based only on psycholinguistic research are bound to be deficient.

Though we have focused our attention on just one of the 25 guidelines in *GDD*, we could have selected a number of others instead, with similar findings. The guideline to "Use the active voice," for example, is violated frequently in the documents we looked at — including the ten plain English ones. Likewise, the prescription to "Write short sentences" is often ignored, as are other prescriptions like "Avoid nouns created from verbs" and "Unstring noun strings." And not only do good writers often ignore these oversimplified maxims, there are many scholarly papers explaining *why* they ignore them.[6] Such explanations, like those put forth in this paper, consider various contextual and rhetorical factors involved in the use of these linguistic forms, not just syntactic factors.

The plain English movement is making important contributions to public communication. The twelve case studies in *How Plain English Works for Business* are a testimonial to the general concept of the movement — and to those writers and editors who make it work. In all of these cases, large corporations have found that adapting their documents to the needs and abilities of their audiences has paid large dividends. But what exactly have the writers of these documents done to achieve such ends? One thing they have not done, judging from our research, is to simply employ the kinds of writer's maxims that the plain English movement has come to be known for. Rather, they seem to have followed many of the practices that good writers in general have followed over the years. If the plain English movement is to fulfill its pedagogical function and teach novice writers how to write comprehensible prose for the general public, it should look more closely at these general principles that all good writers employ. Otherwise, it will be seen as not preaching what it practices.

1. *GDD* provides two other example sentences, but neither is a good illustration of the maxim.

(1') a. We are detailing methods of computing unit costs and requirements for cost detail *which should be* submitted to support your proposed hourly rates.

b. We are detailing methods of computing unit costs and requirements for cost detail submitted to support your proposed hourly rates.

(1") a. This policy does not pay benefits to Park Police *who have been* kicked by horses.

b. This policy does not pay benefits to Park Police kicked by horses.

Sentence 1' is not a legitimate example of whiz-deletion because the modal *should* is not a form of the verb *to be*. Sentence 1" illustrates whiz-deletion but the resulting sentence is not ambiguous. The simple present tense of the main clause verb is not restricted to any particular time-frame (i.e. it is a "general truth"), and thus the inferred tense of the subordinate clause that it governs in (1"b) would also be broad in scope. The two most likely candidates for this role would be the present perfect, as in 1"b, or the simple present, as in (1"a') "This policy does not pay benefits to Park Police *who are* kicked by horses." Most readers, we believe, would not interpret (1"a) and 1"a') as differing significantly in meaning.

2. This ambiguous sentence was found in plain English promotional literature: "We want forms and documents written to be understood by real people."

3. Quirk et al., pp. 937-942.

4. Chomsky and Halle, *The Sound Patterns of English*

5. E.g., Frazier, L., and Fodor, J. D. (1978), "The Sausage Machine: A New Two-Stage Parsing Model," *Cognition 6,* 291-325. Wanner, E. (1980) "The ATN and the Sausage Machine: Which one is Baloney?" *Cognition 8,* 209-225.
6. See, for example Kaufer and Steinberg (1984), McNeill (1966), Walpole (1979), Ohmann (1979).

REFERENCES

Charrow, V., and Charrow, R. (1978) *The Comprehension of Standard Jury Instructions: A Psycholinguistic Approach.* Arlington, VA: Center for Applied Linguistics.

Fodor, J. A., and Garrett, M. F. (1967) "Some Syntactic Determinants of Sentential Complexity," *Perception and Psychophysics,* 1967, *2,* 289-296.

Foss, D.J., and Lynch, R. H., Jr., (1969) "Decision Processes during Sentence Comprehension: Effects of Surface Structure on Decision Times," *Perception and Psychophysics, 5,* 145-148.

Geisler, C., Kaufer, D., and Steinberg, E. (1984) "The Anaphoric Unattended *This:* When Should Writers Use It?" *Written Communication, 2:2,* 129-155.

Hakes, D. T. (1972) "Effects of Reducing Complement Constructions on Sentence Comprehension," *Journal of Verbal Learning and Verbal Behavior, 11,* 278-286.

Hakes, D. T., and Cairns, H. S. (1970) "Sentence Comprehension and Relative Pronouns," *Perception and Psychophysics, 8,* 5-8.

Huckin, T. (1983) "A Cognitive Approach to Readability," in P. Anderson, R. J. Brockmann, and C. R. Miller (Eds.) *New Essays in Technical and Scientific Communication: Research, Theory, and Practice.* Farmingdale, NY: Baywood Press. pp. 90-110.

Kaufer, D., and Steinberg, E. (1986) "Noun Compounds: Four Tests," *Journal of Advanced Composition,* (in press).

McNeill, D. (1966) "Speaking of Space," *Science,* May 13, 875-880.

Moskovit, L. (1983) "When is Broad Reference Clear?" *College Compostion and Communication, 34,* 454-469.

Ohmann, R. (1979) "Use Definite, Specific, Concrete Language," *College English, 41:4,* 390-397.

Quirk, R., Greenbaum, S., Leech, G., and Svartvik, J. (1972) *A Grammar of Contemporary English.* London: Longman.

Selzer, J. (1983) "What Constitutes a 'Readable' Technical Style?" in P. Anderson et al., pp. 71-89.

Steinberg, E. (1986) "A Pox on Pithy Prescriptions," *College Composition and Communication,* (in press).

Walpole, J. R. (1979) "Why Must the Passive Be Damned?" *College Composition and Communication, 30,* October, 251-254.

3

AIDS TO WRITING REVIEW AND REVISION

Can Guidelines Help Writers?

Janice C. Redish and Susan Rosen
American Institutes for Research, Washington, DC

In this paper, we examine the strengths and limitations of using guidelines to improve writing. Experience shows that when we teach guidelines in the context of a model of the writing process, we give writers valuable tools. The Document Design Center's model (a job aid) guides writers to select the appropriate guidelines only after considering the writer's purpose, the audience, and the tasks that the audience will do with the document. Writers must distinguish between guidelines and rules. Although naive writers may follow guidelines slavishly without using common sense, we need not dismiss the practical wisdom and the research results that guidelines embody.

Guidelines allow us to distill research and good practice into chunks of useful advice. Writers can agree that guidelines are useful and yet still argue about the way to state any particular guideline. As we examine a few specific guidelines, we point out that we must use caution in reviewing any corpus of examples from existing documents. The writer may have been constrained for a variety of reasons from producing the document that best uses the writer's own guidelines.

Do guidelines have any place in the toolkit of technical, legal, and business writers? If so, what are their strengths and weaknesses and what role should they play?

We must begin with a definition: A guideline is a suggestion that helps writers achieve the goal of communicating clearly with their readers. Two elements of this definition are crucial.

1. Guidelines are suggestions, not rigid rules or maxims.
2. Writers use guidelines as part of the process of communicating. The choice of guidelines to use depends on many factors, including audience, purpose, and type of document.

GUIDELINES ARE SUGGESTIONS, NOT RULES

We must distinguish and we must help writers to distinguish between guidelines and rules. Guidelines distill research and good practice into chunks of useful advice. Guidelines, however, do not replace the writer's good judgment or the writer's understanding of the writing process. Writers use guidelines to help them carry out their decisions about their purposes, their audiences, and the ways in which people will use their documents.

A writer or a team of writers can develop rules to implement a guideline within a specific context. Guidelines suggest; rules prescribe. Most style guidelines can't be turned into rules. Style guidelines like "Use the active voice" and "Write short sentences" are not inflexible rules. The guidelines don't say "Use the active voice for every verb" or "Make each sentence fifteen words long." Attempts to turn these guidelines into measurable objectives would lead to something like a readability formula. We and others have shown how counterproductive readability formulas are as a way to achieve plain English (Redish and Selzer, 1985; Duffy, 1985; Duffy and Kabance, 1982; Redish, 1981).[1]

Guidelines on format, on the other hand, can usually be turned into rules. For example, in the Document Design Center's handbook, *Guidelines for Document Designers,* we recommend that writers "Use highlighting techniques, but not overuse them" (*Guidelines,* p. 73). In the text, we describe some highlighting techniques and make two important points:

- If you use too many highlighting techniques, you may defeat the purpose of having certain elements stand out.
- If you don't use each highlighting technique consistently, you may confuse your readers.

The guideline by itself, however, does not dictate how to highlight in a particular document. For each document or set of documents, the writers (or designers or managers) should set rules for highlighting. For an insurance policy, we might agree that all headings will be in boldface and that specially defined words will be in italics. On a computer project, we might agree that the names of commands will be in all caps (e.g., PRINT), the names of screens will be in initial caps (e.g., the Main Menu), and what users must type will be in boldface (e.g., Type: **PRINT**), and so on.

Of all the style guidelines that we teach, one of the few that lends itself to becoming a rule is "Be consistent in your choice of words." As a guideline that is good advice for all technical, legal, and business writing. (Note that it is not necessarily an appropriate guideline for fiction, narrative, or essay writing.) In any particular document or project, the writer (or team of writers) has to decide which objects and concepts the guideline will cover and what the rule for each of these will be. The rule might say that we agree, for this project, that a "widget" is always a "widget" and never a "gadget." The rule might say that we agree, for this project, that the verb is always "change" and never "modify" or "update" or "alter." Many computer products suffer because developers and writers didn't create a rule for this guideline.

Rules are project-specific or company-specific. Guidelines have greater generality. The distinction between guidelines and rules is important.[2] It is not made often enough. Unfortunately, the words "style guide" are used both for guidelines and for rules. Some "style guides" are guidelines, that is, general suggestions for clear writing. Others cover the specific rules that writers must use in a particular project. Sometime a "style guide" mixes both guidelines and rules.

The Document Design Center's handbook, *Guidelines for Document Designers,* is a style guide only in the first sense—it contains general suggestions for clear writing. At the Document Design Center, we also create a "style guide" in the second sense for each project that we do. This latter document should really be called "Style rules for Project xxx."

The point of a guideline and the accompanying explanation, examples, related principles, and qualifications is not to prescribe behavior, but to help writers think about what they are doing. Writers have to use judgment in selecting among guidelines and in deciding how and when to implement a guideline. That brings us to the second part of our definition of guidelines.

GUIDELINES ARE PART OF THE PROCESS OF WRITING

The second crucial feature of the definition of guidelines is that guidelines are part of the communication process. The emphasis here is on process. Writers make choices about which guidelines to use, as well as when and how to apply them. These choices depend on the rhetorical situation. Readers' needs and writers' concerns mesh to form an often complicated communication process. Guidelines can act as a path through all of the variables of the reader-writer relationship.

What choices do writers make?

Guidelines have to be understood within the context of a complex process in which writers move back and forth among the phases of the process, planning, drafting, evaluating, and redrafting. At the Document Design Center, in all of our work and our workshops, we use and teach guidelines in the context of a model of the writing process that urges writers to think about their own purposes, their readers, how the readers will approach the document, and other contextual features.[3]

When we wrote *Guidelines for Document Designers,* we were working with professionals—mostly lawyers—who had traditionally given little thought to their lay readers or to the fact that their documents were used for looking up information rather than for reading from cover to cover. For them (and for our later clients, including financial aid officers, software developers, medical doctors, insurance specialists, forms designers, and engineers), we found that a prewriting analysis of their goals, their audiences, and the ways that people would use their documents was essential to their being able to choose guidelines wisely.

For that reason, we wrote in the introduction to *Guidelines:*

> The guidelines in this book come into play only when you are ready to organize and write your first draft. Before you get to that stage, you should have done a careful analysis of your task, your purpose, your audience, and the task you are asking of your readers (*Guidelines,* p. 3).

Our assumption in writing *Guidelines* was that writers would use judgment in selecting and applying the guidelines; and, in our experience, that is what writers do. We are trying to hone their judgment, to broaden their views of acceptable guidelines, to show them, by example, the effects of applying different guidelines.

Guidelines are aids to the writer, they are not meant to supplant human judgment. We see little value in programs that computerize guidelines and allow no changes in style for different contexts. Some programs turn guidelines into unchangeable rules. For example, when it first came out, one computer program marked every sentence over twenty-two words in length as a "long sentence." This program created an arbitrary rule from what should be a flexible guideline. Even if users could set a different number for "long sentence" in each document, depending on the rhetorical situation, no single number would be appropriate for all the sentences in a document. While it might be useful for a writer to reconsider all sentences that are longer than a specified number of words, the writer must ultimately make the decisions about revising each sentence that the program marks.

Furthermore, in deciding whether (and how) to revise each sentence, the writer must consider not only the rhetorical situation (purpose, audience, reader's task), but also the broader discourse in which the sentence occurs. In *Guidelines,* therefore, we included under most guidelines "Qualifications" that caution the writer to attend to the discourse context. Under the guideline "Write short sentences," for example, we remind writers that "a well-constructed long sentence can give needed variety to the rhythm of the passage. Also, some ideas may make more sense when you present them together in one long sentence than in several short ones."

Can writers do without guidelines? Actually, writers already do use guidelines—either explicit guidelines or ones they have internalized so much that they use the guidelines automatically. Guidelines are a necessary part of any heuristic. Whether writers use the Document Design Center's process model, Tom Duffy's model (Duffy, 1981), Paul Anderson's model (Anderson, 1984), or follow the heuristics in Linda Flower's *Problem Solving Strategies for Writing,* (Flower, 1981), at some point, they use guidelines.

Who uses explicit guidelines and how do they use them?

In selecting the topics to cover in *Guidelines,* we were particularly conscious of our own audience and the types of documents that they write. Our primary audience has always been people who write a lot in their jobs, but who are not trained writers. These people are professionals who work in business, government, law, medicine, and related fields. They are well-educated, articulate speakers of English who already write adequately in their work. However, they want to write better. This is the same audience that the authors of most "how to" writing books are addressing.

In light of recent criticisms of guidelines, we were curious about who uses guidelines and how they use them. To find out, we interviewed thirty people. Twelve were experienced professional writers, eight were recent graduates of technical writing programs, and ten were professionals in fields other than writing. In informal individual interviews, we asked all thirty whether they use guidelines in their writing, and if so, how they use them.

Most of the professional writers said that they do not use guidelines. They "know" how to write to different audiences in different situations. Many of these writers, however, felt that guidelines were necessary and reported using them earlier in their careers though they no longer needed to. Clearly, the professional writers in our sample believe that they have internalized a set of guidelines and heuristics for when and how to apply them.

The recent graduates had mixed feelings about using guidelines. Some felt that guidelines helped them stay on task, serving as continual reminders of the general style and tone of writing that they were trying to accomplish. Others dismissed guidelines as information that they had already learned in school.

The people whom we might call "professionals who write" were the strongest advocates of explicit guidelines. These people are all well educated in areas other than writing and all have to write in their jobs—to communicate with others in their company, with other specialists in their field, with people in related fields, and with customers.

Nine of the ten "professionals who write" reported that they rely on guidelines in their writing. Six use guidelines to get started and stay organized. Two use books of guidelines for examples of good style. One woman in this group said that guidelines were a "lifeline," and that without them she felt lost and unsure of how to proceed.

Our small survey confirmed the hypothesis that we had when we developed *Guidelines for Document Designers*. Professionals in disciplines other than writing who have to write a lot in their jobs are grateful for the practical advice of style guides like *Guidelines*.

Other studies have also shown how useful guidelines can be. Barry Jereb in this volume describes how Allen-Bradley Company has greatly improved its technical manuals by changing the page layout and the writing style (Jereb, 1986 and this volume). In support of this new style, the Communications Design Center of Carnegie Mellon University developed for Allen-Bradley a draft of a manual entitled *Guidelines for Preparing Technical Documentation*. Allen-Bradley also gives a four-page set of guidelines to companies that write for Allen-Bradley. These guidelines and the presentation of them in terms of readers' problems are taken directly from the Document Design Center's course book, *Writing in the Professions* (Goswami, Redish, Felker, & Siegel, 1981, pp. 110—11).

WHAT MORE DO WE NEED TO LEARN ABOUT GUIDELINES?

Since writers must use judgment in applying guidelines, many writers need help in knowing when and how to use them. One fruitful area for further research is to elaborate on the guidelines in order to help writers know how to use them appropriately.

Kaufer and Steinberg's recent study, "On Revising Noun Compounds: Four Tests," is exactly the type of elaboration that we need. Noun compounds, which are sometimes called noun strings or noun sandwiches, occur when a writer modifies a noun with one or more other nouns. For example:

telephone receiver
day care center
health care claim summary
host area crisis shelter program planning workbook

As Kaufer and Steinberg point out, some of these noun compounds don't cause problems for readers while others do. How can a writer know when to untangle a noun compound for readers?

Kaufer and Steinberg propose four tests for writers to apply and suggest that any noun compound that fails even one of these tests should be revised. Kaufer and Steinberg's tests "bid writers ask themselves whether the compound (1) falls within the reader's knowledge, (2) functions as common usage, (3) packages ideas efficiently, and (4) is stylistically efficacious." As they point out, their tests mean that "to know when to revise a noun compound, writers must know something about its 'intrinsic' properties. . . . But the most important knowledge they will need is 'extrinsic'—*who* must understand the compound and to what purpose?" A study like Kaufer and Steinberg's helps writers understand how to use guidelines. Similar explorations of other guidelines would be welcome.

In addition to thoughtful pieces that elaborate on the guidelines, we need research to help writers know which guidelines to apply and how to apply them in different circumstances. As far back as 1980, Redish was calling for research that would validate or challenge specific sets of guidelines in specific situations. Because guidelines are meant to help writers produce documents that readers can easily understand and use, research on guidelines must involve testing documents with readers—and with readers who truly represent the audience for the document.

Unfortunately, guidelines are now being challenged without that type of research. In two recent studies, Huckin, Curtin, and Graham (1986 and this volume) and Steinberg (1986) have examined style books and have raised concerns that writers may take guidelines as pithy prescriptions and use them without the appropriate thought and judgment.

Huckin, Curtin, and Graham, for example, take issue with the Document Design Center's guideline, "Avoid whiz-deletions." "Whiz" stands for "which is." Whiz-deletions occur in subordinate clauses, usually truncated passives, when the relative pronoun and verb are not explicitly stated in the sentence.

The authors suggest that we can test the appropriateness of the guideline by seeing whether or not expert writers use it. They select a corpus of "plain English" documents and compare the number of whiz appearances with the number of whiz-deletions. They find that sentences with whizzes deleted are more common than sentences where the relative pronoun and verb appear. They also argue that many sentences with whiz-deletions are clearer and more appropriate than the same sentence would be with the relative pronoun and verb.

We must treat this methodology with caution. Not every sentence in a plain English document is in plain English. Usability testing and protocol-based revision have taught us that even expert writers sometimes write sentences that readers don't understand. The test of guidelines should not be whether good writers use them but whether they produce documents that readers can understand.

Moreover, the actual document may not represent the writer's expert judgment in every sentence. Real-world documents are compromises. The final document probably went through several rounds of negotiations with lawyers, technical specialists, managers, and other reviewers. In the end, the writer may have had to make changes that violate the writer's guidelines.

Counting whiz-appearances and whiz-deletions in "plain English" documents also misses another important point. We wrote *Guidelines* to help people analyze and rewrite poor documents. We were helping people to understand and rewrite sentences like this one:

> Students receiving Basic Grants based on incorrect information submitted on the application will have to repay any funds received.

The audience for this sentence is high school seniors from economically poor families and their parents, who may have less than a high school education. The writer's purpose is to warn these students about the consequences of cheating.

Many "professionals who write" don't realize that this one sentence combines five clauses and that three of the subordinate clauses are imbedded in the main clause. Putting the whizzes back makes the complexity of the sentence stand out.

> Students *who are* receiving Basic Grants *that are* based on incorrect information *that they* submitted on the application will have to repay any funds *that they have* received.

As soon as the writers of this sentence put the whizzes back in, they realized that it was too long and cumbersome. With our help, they threw it out and replaced it with an entirely different sentence—one based on the scenario principle and several of the other guidelines in *Guidelines*. The new sentence is:

> If you get a Basic Grant by giving incorrect information, you will have to pay it back.

A scholar who looked only at the revised version of this document would miss the value that our guideline on whiz-deletions had for the people who were revising that poor document.

We wrote *Guidelines* for a particular audience—writers of public documents. We didn't write *Guidelines* for students, although many university teachers have found that *Guidelines* works well when presented in the context of a process model or as part of a course that is built on problem-solving heuristics. In other style guides, for other groups, we have stressed different guidelines.

Huckin, Curtin, and Graham rewrite the guideline on whiz-deletions to have a different emphasis from the one that they read in *Guidelines*. Their rewrite does not trouble us. Throughout *Guidelines,* we tried to raise writers' awareness of aspects of sentence structure that they may have paid little attention to in their earlier work. Today, we would probably rewrite the few pages on this guideline to say something like this: Watch out for whiz-deletions. They sometimes cause ambiguity. If you have a long, complex sentence, the problem may be whiz-deletions. After you put the whizzes back in, consider rewriting the sentence.

Other aspects of Huckin, Curtin, and Graham's article do trouble us. They read the guideline as a prescription to be applied to all sentences in all documents for all readers, and they worry that writers will take the guideline as a maxim to be followed blindly. Huckin, Curtin, and Graham also complain that the guideline on whiz-deletions seems to contradict the guideline to avoid unnecessary words. If the relative pronoun and verb clarify an otherwise ambiguous sentence, however, they are necessary words. In this case, there is no contradiction.

Huckin, Curtin, and Graham's complaint, however, reveals a deeper misunderstanding of guidelines. Guidelines do in fact conflict with each other. Only rules are rigid; guidelines require judgment. For example, unstringing a noun string usually makes a sentence longer. Replacing a nominalization with a subject and action verb often does the same. Yet, all three guidelines, "Unstring noun strings," "Use action verbs," and "Write short sentences," are valid and important guidelines.

Steinberg (1986) is also concerned with guidelines as prescriptions when he takes the authors of a book on technical writing to task for advocating "short, precise sentences" when they do not explain how short a sentence should be. He then examines the average sentence length in parts of the book to see if he can find out in that way what the authors mean by a short sentence. He finds that the average sentence length for these parts of the book ranges from 13.5 to 30.8 words. Rather than conclude that this is what the authors mean by a short sentence, he concludes that they are not following their own guideline.

That's not the only possible conclusion, however. We could also conclude that the book does exemplify what the authors mean by short sentences. Our conclusion would be that short sentences don't have to be only 13 or 18 words long. The authors might be saying, by example, that sentences should average about 25 words over the course of the document. For a book of readings on technical writing that is meant for college students, 25 words may be an appropriate average sentence length.

Steinberg's fear is that "many—even most—readers of a book on style would think that an instruction to keep sentences 'short' would mean that 13 was better than 18 for an average sentence length." That's not our experience. In technical and legal documents, we are trying to get writers to think about the burden that a sentence of 60 or 80 or 150 words puts on the reader. To our writers, 20, 25, 30, or even 35 words would make a sentence short.

The pithy expression of a guideline may seem to be a simplified maxim, but the authors of practical advice on style don't expect writers to take each guideline literally for all documents and all situations. In *Guidelines,* for example, each guideline includes the heading, the explanation, examples, related principles, and qualifications. In the book that Steinberg considers, the sentence about short, precise sentences is in a context of many sentences on the same page that say that writers must make choices and use judgment in their writing.

The fear that writers will use guidelines as maxims to be followed blindly is not borne out in our experiences with people who write on their jobs. For example, the first guideline on writing in *Guidelines* is "Use the active voice," which is an important suggestion for technical and bureaucratic writers for whom passive has been the traditional model. We don't advocate using active all the time. In the qualifications to the guideline, we give several examples of situations in which a passive sentence would be appropriate. But even writers who only see the maxim stated in a checklist without our qualifications don't write only active sentences. They understand that the guideline is helping them to change the overall voice of their document, not every sentence in it.

Guidelines serve a very useful purpose for writers, particularly for people who are not trained as writers but who write on their jobs. Guidelines encapsulate advice to stimulate writers to think about particular aspects of their work.

We thank Dr. Edward F. Redish of the University of Maryland and our colleagues, Carolyn Boccella Bagin, Joanne Landesman, and Sarah Spatt, for fruitful discussions and for their perceptive comments on earlier drafts of this paper.

1. Duffy and Kabance conducted a study in which navy recruits read and answered questions on passages that were 1) unchanged, 2) changed to have shorter sentences, 3) changed to have shorter words, 4) changed to have both shorter sentences and shorter words. The changed passages improved comprehension in only one condition. Duffy and Kabance conclude that "readability formulas and guidelines are inadequate as a means of producing more understandable text."

We should not read this sentence as a condemnation of guidelines in general. Duffy and Kabance used only two guidelines in revising their texts—"Use shorter sentences" and "Use shorter words." Their study only showed that these two guidelines by themselves are not sufficient to make text easier for readers to understand.

2. Dr. Joseph Dumas of AIR has a particularly clear description of the differences between guidelines and rules in his book, *Designing User Interfaces for Software,* Prentice-Hall, 1988.

3. See Battison and Goswami, 1981, Goswami et al., 1981, or Redish, Felker, and Rose, 1981, for a picture and explanation of the model. Note that we present the model linearly for two reasons: It works very well that way as a job aid for writers; and it makes writers see the importance of evaluating their document, not only internally while they are writing, but externally, by testing it with readers. In reality, however, the writing process is much more complex than the way we show it in the model as a

job aid. The research of Flower and Hayes (for example, Flower and Hayes, 1981) has shown many of the complexities of the process.

REFERENCES

Anderson, Paul, "What Technical and Scientific Communicators Do: A Comprehensive Model for Developing Academic Programs," *IEEE Transactions on Professional Communication,* September 1984, Vol. 27, No. 3, pp. 161–67.

Battison, R., and Goswami, D., "Clear Writing Today," *Journal of Business Communication,* Fall 1981, Vol. 4, pp. 5–16.

Duffy, Thomas M., "Organising and Utilizing Document Design Options," *Information Design Journal,* 1981, Vol. 2/3&4, pp. 256–66.

Duffy, T. M., and Kabance, P., "Testing a Readable Writing Approach," *Journal of Educational Psychology,* October 1974, Vol. 74, pp. 733–48.

Felker et al., *Guidelines for Document Designers,* American Institutes for Research, Washington, D.C., November 1981.

Flower, Linda, *Problem Solving Strategies for Writing,* Harcourt Brace Jovanovich, Inc., New York, NY, 1981.

Flower, Linda, and Hayes, J. R., "A Cognitive Process Theory of Writing," *College Composition and Communication,* December 1981, Vol. 32, pp. 365–87.

Goswami et al., *Writing in the Professions,* American Institutes for Research, Washington, D.C., November 1981.

Huckin, Thomas N.; Curtin, Elizabeth H.; and Graham, Debra, "Prescriptive Linguistics and Plain English: The Case of 'Whiz-deletions'," *Visible Language,* Spring 1986, Vol. 20, pp. 174–87. Reprinted in this volume.

Jereb, Barry, "Plain English on the Plant Floor," *Visible Language,* Spring 1986, Vol. 20, pp. 219–25. Reprinted in this volume.

Kaufer, D., and Steinberg, E., "On Revising Noun Compounds: Four Tests," *Journal of Advanced Composition,* 1984, Vol. 5, pp. 51–64.

Redish, Janice C., "Research in the Junior Year Writing Course," *Proceedings of the Inaugural Conference of the Maryland Junior Writing Program,* College Park, MD: University of Maryland, Department of English, 1980. pp. 159–67.

Redish, Janice C., "Understanding the Limitations of Readability Formulas," *IEEE Transactions on Professional Communication,* PC-24 (1), 1981, pp. 46–48.

Redish, Janice C.; Felker, Daniel B.; and Rose, Andrew M., "Evaluating the Effects of Document Design Principles," *Information Design Journal,* 1981, Vol. 2/3&4, pp. 236–43.

Redish, J., and Selzer, J., "The Place of Readability Formulas in Technical Communication," *Technical Communication,* Fourth Quarter 1985, pp. 46–52.

Steinberg, Erwin, "A Pox on Pithy Prescriptions," *College Communication and Composition,* February 1986, Vol. 37, No. 1, pp. 96–100.

Computers for Composition: A Stage Model Approach to Helping

Lance A. Miller
IBM T. J. Watson Research Center, Yorktown Heights, New York

This paper is concerned with how computers can assist text composition. It is a review of what has been or is being done — particularly in the EPISTLE project at the Yorktown IBM Watson Research Center — and it is also a preview of what could be done in the future. The discussion is centered around a stage model of composition adapted from computer programming. The model's psychological validity, although credible, is secondary to the rather natural way it appears to organize possible computer function supporting composition. Some stages of this model clearly require powerful language technologies, and the stronger interest is clearly in these. Nevertheless, a number of functions are identified for other stages, which functions do not require extensive linguistic capability to implement or enhance. Throughout the paper there is a deliberate bias towards "practical" writing — writing whose fruits are of unquestioned value and whose authors receive monies undetermined by word counts or reviewers' praise. The notion of this type of writing — most frequent by far — is that the composition meets a requirement of business or social commerce, whether it is an interagency status report, product documentation, a step in the process of attempting to free a client, or a worded advertising fantasy designed to attract cash customers. The paper concludes with some consideration of specialized audiences for the various computer functions as well as the roles educators might play in promoting (or obstructing) their development.

A STAGE MODEL OF COMPOSITION

In this paper "composition" means a cohesive body of text expressed primarily in full grammatical constructions, with at least one primary communication objective, and usually embodied as a recognizable writing genre (such as a user manual, memorandum, or obituary). A composition is also the desired end product of the process of composition. While there are many important comments and observations on this activity (e.g., Flower 1981; Flower & Hayes 1980; Mathews & Stevenson 1976), there is certainly room for additional research, particularly as concerns detailed cognitive models. The idea of a "software-design" model applied to composing is one attractive possible extension.

The flavor of the software-design stage model is to identify a variety of stages for programming activities and to propose various tools and environ-

Reprinted by permission of the publisher, *Visible Language*, Rhode Island School of Design, Providence, RI 02903.

ments to support each stage as well as the overall task (e.g., Yourdan & Constantine 1979; Orr 1977; Miller 1978). Although the stages are expressed as a linear sequence, a considerable amount of iteration among various stages is presumed throughout. Similarly, while the end-product appears to be better when all of the stages are passed through, in order, the model never purports to describe more than what some people do some of the time. A simplified form of the model is shown in Figure 1.[1] This simplified software development stage-model can be adapted almost directly to the process of composing by substituting some terms, as shown in Figure 2.

In stage 1 the problem to be solved by an eventual composition is considered and elaborated into a set of requirements which must pay at least some heed to the three primary rhetorical factors of audience (and speaker), topic, and communication purpose (e.g., Mills 1952). The additional rhetorical factor of the nature of construction of the actual message is not an aspect of this stage but of the next (or later).[2]

In stage 2 the initial decisions made for audience, topic, and pragmatic objectives are to be taken together to determine two emerging textual features: (1) the selection of the appropriate genre-format, and (2) the initial choice of style.[3] The first notion, genre-format, combines the selection of a particular writing genre (usually domain and intent specific — a "missive," a "persuasive argument") with a conventional schema for sequencing and emphasizing the parts thereof (a "business letter," a "court of appeals brief"). The second notion of style choice is taken here narrowly to mean the "strategy for communicating information," and it can operate both as a selector of genre-formats and as an organizer of material within a particular format.

1. Functional requirements
2. High-level design
3. Detailed design
4. Coding
5. Testing
6. Installation
7. Modification

Figure 1. Software production stage model

1. Communication requirements
2. High-level plan
3. Detailed plan
4. Scribing
5. Critiquing
6. Release
7. Adaption

Figure 2. Text composition stage model

The detailed structuring of a composition in stage 3 involves considerable fleshing out and elaborating facts, arguments, and themes — though still in pre-discourse form. Here more than elsewhere the items from unorganized lists ("points to make," "ground to cover," "essential information") get transferred into a detailed sequence, and the possibilities and necessities of coherence begin to become known.

If the process of composition is likened to a drama, then the theatrical climax would occur in the 4th stage, *scribing*. Up to this point, the planning and evaluation for composition could be quite similar to that for painting, writing music, or choreographing. Here, the words used as symbols for concepts of communication in extra-linguistic fashion now become actual text; the illocutionary intent is now instantiated as actual locution, capable now of that most dreaded of readers' acts — being quoted! No longer can a word or phrase merely stand for the intended sense, but *le mot juste* must be found; government, binding, and overall predication must be accomplished, with due consideration to all of the complexities of movement, raising, coordination, and agreement — and this just for a single sentence! How fearsome the choice then as inter-sentential organization must be achieved: here the strong marker of discourse structure ("First of all, . . ."), there the repeated theme by hint of lexical collocation (". . . the accident . . . the damage . . ."), and now the eschewing of anaphoric "he" for cataphoric ". . . these three:" Level after level must coordinate and cohere, topic to sentence, sentence to paragraph, paragraph to chapter — and all to each and every word. No wonder is it then that the slightest hesitation, the merest stickiness of fit, the briefest delay of word-finding all can bring such feverish (or labored) activity to an instant (or grudging) halt — to reconsider the last word, the format appropriateness, or even the whole stage 1 approach.

Such examinations of text, whether planned or forced, bring the process, however momentarily, into the 5th stage, *of critiquing*. Included are issues as mundane as the number of 'r's in "embarrass" or as complex as whether the main point is really being made. This stage is but a springboard, mostly, for revisiting previous stages. In particular, the visiting of the scribing stage after critiquing is called revision.[4]

The 6th stage, *release,* identifies the point at which the text leaves its maker and has an opportunity for a life of its own. Publishing of the text is one of the strongest forms of release (in several of its meanings!) although in general this occurs just whenever the text is truly "given away".

The last, the 7th stage, *adaptation,* refers to the modification of a document to reflect new requirements, new information to be included (updates), or error correction. The adaptation of previously released text almost always involves a new release of some sort — printing, copyright, or even published "errata." However, this stage is not an important aspect for the present focus.[5]

Stages to be discussed

The last two stages, while still important aspects of the composition process, go very far afield from the present topic of personal computer aids to writers. It's quite likely that the emerging Expert Systems technology could, some day, be applied to these aspects, but that speculation is outside of this paper. The other stages are discussed roughly in terms of the extent to which computers presently support them. Thus, the fourth stage of scribing is discussed first, then stage 5 critiquing, and then stages 1-3.

COMPUTERS AND SCRIBING

Computers have had the most obvious impact in the scribing stage — the actual generation of text within some medium. Clearly, the most important role has been played by the word-processor, the first topic in this section. However, the role of formatters is also extremely important, and, finally, there are surely some number of yet unthought-of useful scribing aids.

Word-processors

Computer word-processors — and, lately, electronic typewriters with similar features — have revolutionized the process of business and technical writing. Perhaps the most glowing praise concerns not the actual transcription of words, but their editing — movement of text elements from one part to another, global changes, the location and modification of particular targets, and any of a host of other similar specialized functions (e.g., Sekuler 1985). These text-modification activities are still considered to be part of the scribing stage — particularly since they involve the computer editor; however, the thinking and evaluations that usually precede them are said to be of the following stage, critiquing.

The impact of computer word-processing on commercial scribing is all the more remarkable when it is considered that the computer software has no real idea of what the words mean or what the text is about. The functions operate on the notion of "token" — character strings delineated by blanks (or perhaps other special characters) — and the notion of "line record", a single line of the edited file as wide as permitted by the editor at that time. However, the editing capability provided by this primitive token analysis is quite adequate for most functions, and many users are not even aware that the editors do not share their view of word or sentence.[6]

It is not only the professional writers who have been influenced by this word-processor aspect of computers. There is also abundant evidence that word processing is welcomed eagerly by students in college and elsewhere, as the word-processing capability is introduced. There is a serious question for many, so we are told, whether they could ever write with pencil and paper again!

Computers used for the scribing phase are really inseparable from the question of the features available on the computer editor, and this has been an area of great interest to researchers, both from the standpoint of designing better editors (Miller & Thomas 1977; Embley & Nagy 1981; Roberts 1980) and from the point of view of editing as a psychological task (e.g., Card et al. 1976). The very large number of computer editors that are available — even from a single manufacturer — is indicative of the extent to which this area of the relation between computers and composition is being explored — and exploited.

Formatting

With pencil and paper and older typewriter technologies the act of scribing also — simultaneously — involved the act of deciding how to format the text on the page: whether to center, skip a line, indent, etc., for the grosser non-alphabetic manipulations; and, for the text itself, whether to capitalize, underline, change colors, or even change fonts. Thus, if a word was to be centered, capitalized, and underlined (to serve as a topic heading, for example), then these activities co-occurred with the actual generation of the characters of that word and the positioning of them on the paper.

Nowadays, with word-processors, you can defer the actual formatting until a later time by embedding special formatting codes in your text in special ways, and then sometime later sending your text file with these codes to a special formatting program to produce hard-copy or soft which has the text formatted according to your specifications; some of the formatting languages are SCRIBE, WATERLOO SCRIPT, and IBM SCRIPTVS. And, around the corner, are so-called WYSIWYG editors (pronounced "wiz-ee-wig" for "What You See Is What You Get") in which your typed formatting code is immediately executed, so that the text is formatted directly without a subsequent step. Nevertheless, since formatting was traditionally an integral part of the scribing process, and since formatting considerations still are very much a part of scribing, in terms of the requirement to embed formatting codes within the text being scribed, this activity is discussed here — and not, as it might well be, as part of the critiquing process.

Formatting technology has been around for a long time but is recently undergoing some important pervasive changes. Publishers are very near to agreeing to a standard for "marking up" documents in terms of classifying the contents of a particular body of text with certain "tags" preceding (and often ending) the text. This standard is being called Standard Generalized Markup Language or *SGML*. Thus, there are tags which identify the beginning of a document title, a footnote, a chapter heading, or an index. There are also tags which identify any arbitrary phrase, within any arbitrary document section, as, say, a definition, a citation, or simply something to be highlighted. For example, the reference to ". . . or *SGML* . . ." above (and here) was typed exactly as follows:

or :cit. SGML:ecit.

such that the acronym SGML was surrounded by two tags, each beginning with a ":" and ending with a "." — the tag itself being "cit" for citation. Similarly, the first three numbered lines of Figure 1 were produced by the following set of formatting control tags embedded with the text:

:ol.
:li.Functional requirements
:li.High-level design
:li.Detailed design

The :ol. tag indicates that an ordered (numbered) list is to be started, and the :li. tag indicate "list-items" to be put one under another and numbered consecutively (see the IBM manual on the Generalized Markup Language, 1984).

These tags have two important advantages over first-generation formatting control words. First, they are nouns rather than verbs: they describe the nature of the text contained rather than specify what particular line-skipping, indenting, or other actions should take place. Thus, in the older IBM formatting language of SCRIPT, the control word .sk could be used to skip a line and start a new paragraph, without indentation; in the newer GML tag language, :p. is used instead. This symbol means that "the following is a new paragraph," but the actual interpretation is not specified by the author but by conventions observed at the local installation. In some situations, then, :p. could be interpreted as "skip a line and don't indent"; in another situation it could mean "skip and indent five spaces"; and in still a third — a publisher of children's stories, say — it could lead to that stylistic ballooning or enlarging of the first letter of the first new paragraph word. The author is therefore freed from having to think about ultimate formatting arrangements (see Misek-Falkoff 1983, for detailed discussion of these points).

A second advantage of the use of "noun" tags is that these descriptors can imply formatting manipulations far more numerous and complex than simply skipping a line and indenting. For example, ordered list items don't have to be numbered by the author; they are simply typed in with the :li. tag and can be re-ordered or otherwise manipulated, and the formatter will number them automatically. Similarly a large complex document with multiply embedded, and numbered, sections (e.g., "1.0 2.0 2.1, 2.11, 2.12, 2.121, 2.1211, . . .") can be outlined using certain heading tags which do not force the author to think about numbering or other matters at all;[7] when formatted, the headings are not only numbered correctly, but the text associated with the heading is also printed in the locally-preferred combinations of font-styles, font-sizes, and bold-face vs. plain.

The tags therefore provide means for freeing the user from thinking about how to format documents while generating the text, and there are many new schemes for simplifying the learning and use of such tags. The tags could also be considered to be structural landmarks in the text, and they might well assist authors by providing cues concerning the organization of the document. In fact, authors might even use tags — particularly heading tags — to outline a part of the document first and then fill in the section with text. As is discussed in the later section on planning, this idea is extended to provide for the computer to produce fully formed outlines for the author to follow, based upon the author's selection of a particular genre.

Additional aids

There are extraordinary opportunities for clever programmers, on the one hand, and educators with a feeling for exploiting computer technology, on the other, to recognize and act on the possibilities for new computer function for the scribing process. Below are a list of a few ideas (which can be found in some existing software in some cases):

Expanders. These can make it easier for people to refer to long or complicated phrases or character expressions as they work along, via short strings: thus .ibm., when entered, might cause the system to expand and enter the string to be "International Business Machines Corporation."

Phras-ers. This is the idea of breaking text into smaller meaningful units. If a person entered several sentences at one time, the code could intercept the keyboard input, find the ends of sentences, and put each sentence on a separate line. This code might also break sentences into various kinds of phrases, using simple phrase-finding algorithms, such that the subject of a sentence was on one line, the verb elements on another, and the direct object information on a third. These types of text separations might help an author see non-felicitous constructions or word-choices, or it might help in evaluating sentence complexity.

Revision marking. When portions of a text have been changed in a session (or previously), the altered sections could be indicated by changes in displayed color, margin indicators, or italics, etc. This might well assist an author in picking up the thread of a paper being worked on, and it could provide a useful record of editing activities.

On-line dictionaries. Many authors interrupt their writing to look up a word, to check its meaning or to discover a synonym or other type of substitute. It would be very nice to be able to access a dictionary just by putting the cursor on a word in the text and pressing some function key to have the definition appear on the screen.

On-line text/data retrieval. How often it happens (at least to this author) that in the midst of writing something one realizes one needs a name, or some piece of data, or some other example text to look at (and copy). How nice it would be, then, to be able to open, say, another window on the screen, peruse various other files until one finds what one needs, and then — presto — push the right buttons and have that information transferred into one's working file.

Speech input/output. Perhaps continuous speech recognizers won't be available for some while yet, but even small vocabulary isolated word recognizers would be very useful as input devices for text-editing commands and formatting controls. On the output side, thinking how useful it is often to read one's paper out loud, or to have it read to one, a high quality voice synthesized speech reading of one's text might be very nice.

COMPUTERS AND CRITIQUING

The second stage of composition that computers have entered extensively is that of critiquing, the stage following scribing. It is in this area, for the first time, that linguistic notions begin to be introduced into the computer software. While the degree of critiquing function that is now operational is quite broad, the possibilities are even greater, as indicated in the following section.

The space of possible critiques

The act of critiquing involves the examination of something relative to a set of standards followed by a part of the findings. In order to appreciate the rather vast possibilities of critiquing for compositions, five distinctions can be made: (1) the examination text-unit, (2) the report text-unit, (3) the critique type, (4) the strength of the critique report, and (5) the linguistic perspective of the critique. The levels of text we find generally useful to identify are: character, morpheme, word, token, phrase, clause, sentence, paragraph, section and document. A morpheme might occur as a stand-alone syllable in the text if, for example, words were hyphenated at the ends of lines; a token may be a word alone or a word combined with punctuation or other non-alphabetical characters (e.g., "and/or"), or it may be a word combined with a special formatting code, like the markup tags (e.g., ":cit.SGML:ecit."); clauses are identified as such only when there is more than one of them in a sentence; a section is a major portion of a document — like chapters in a book — and a document is the entire composition, however long, whatever the genre.

The examination text-unit refers to the unit of text which is examined for the presence of some target. If the critique is that of spelling-checking, then the examination text-unit is a word; if the critique is to determine whether numbers were used (instead of spelling quantities out in words) the unit is a character.

The report text-unit is the unit at which the critique report is made, and this unit is either the same as the examination unit or else larger. An example of the latter instance is when a text is critiqued for low frequency words (examination-unit = word) and the results are summarized on a paragraph basis (report-unit = paragraph), e.g.: "This paragraphs contains the following low frequency words . . .." Situations in which the report text-unit is larger than the examination text-unit are called aggregate critiquing.

The third distinction, critique type, refers to the manner in which the critique is made, and the two options are isolated vs. relative. In an isolated critique a particular examination-unit is compared against a standard, and the judgment can be rendered without taking into account the characteristics of that unit relative to other units. Thus, checking for spelling errors, incorrect capitalization, overly-long sentences, etc., all involve an isolated critique. In contrast, a relative critique checks the characteristics of one text-unit (having certain features) against the characteristics of another text-unit (having different features); the logic of the comparison is along the lines of "if the first unit has an aspect X, then the second unit must have an aspect Y." Most ungrammaticalities, such as disagreement in number between subject and verb, involve such a relative type of critique.

The fourth distinction concerns critique strength for which there are also two possibilities: right-wrong vs. threshold. A right-wrong judgment is one in which one can say "Right!" or "Wrong!" without fear of contradiction (from experts), as in the case of the majority of grammatical errors. Thus, "two apple" and "He aren't" are just plain wrong. On the other hand questions of style are not only matters of taste but also the province of the organizational enterprise authorizing the composition; one must be very careful in critiquing these things, since observations like "the sentence is very long" or "the structure is not parallel" aren't sufficient for strong criticism in themselves. Stylistic problems must typically needs be reported with some deference and sensitivity to the fact that the author and critiquer may not share the same standards. One means of systematically handling the problem of varying stylistic standards is to arrange to have each stylistic evaluation result in the computation of a single number whose value grows with the severity of that particular gaffe; this value can then be compared against the threshold for a particular enterprise, and, if it exceeds that threshold, a suitable commentary is provided.

The fifth factor, the linguistic perspective, invokes the system of semiotic theory (cf. Morris 1946), which identifies three levels of linguistic consideration: syntactic, semantic, and pragmatic.

Such a large multi-dimensional critique space is rather unwieldy in practice, especially since the development of language technologies is really in its infancy. Accordingly, we have organized the EPISTLE critiquing function into three types: word, grammar, and style. Word level critiquing now includes consideration of syntactic problems (misspellings) and someday will be extended to semantic problems (e.g., inappropriate word-choices — using "infer" where "imply" is intended). Grammar critiquing focuses on sentence-level phenomena, and style can be discussed from the word-level up through the whole document.

Word critiquing

The most prolific area of critiquing is that of words, especially for detecting errors in spelling. Recently, some linguistic competence has been introduced into the computer software, in the form of some knowledge about "words" — although this knowledge may be quite primitive and consist only of a list of character strings: anything found to be on the list is considered to be a word; anything not found on the list is a spelling "error." Despite the low level of linguistic knowledge about words possessed by computer spell-checkers, there are some remarkable successes, and the function delivered is surely valuable (e.g., Durham et. al. 1983; Pollock & Zamora 1984). The function has been augmented in other products, including IBM's PROOF product, to include detection of awkward phrases and the capability to suggest synonyms (e.g., Leerburger 1981; IBM 1983).

The most extensive dictionary capability function appears to be that provided by the IBM Yorktown Watson Research Center's experimental EPISTLE system, whose on-line dictionary provides strings of syntactic and semantic features, for use by the grammar component, for some 130K words (cf. Byrd 1983, 1984; Byrd and McCord 1985).[8] Perhaps half again as many words are recognizable by use of the extensive capability of this system to analyze an unknown word into and deduce its part of speech and other characteristics (so-called "inflectional and derivational morphology").

Word critiques which involve reporting of word problems at the sentence level include (1) incorrect spellings, (2) non-preferred spellings, (3) incorrect abbreviations, and (4) incorrect capitalization (e.g., "dr."). These four sentence-level critiques are also critiques which are triggered by the simple presence of the problem — occurrence-level targets. A sentence-level examination critique involving an aggregate-level report is (5) one which checks for repetitions of substantive words in the same sentence (e.g., "The cool light was cool on my face.").

Other possibilities for word-critiquing at the discourse level include five main classes: (1) observations on the relative frequency of word usage from various types of dictionary sources, (2) observations on tonal features of the words, (3) detection of hackneyed words or phrases, (4) detection of incorrect or inappropriate word usage, and (5) comments on relatively subtle changes in usage of words, particularly the same words, from one part of the text to another. The first two categories involve aggregate reports on the text, still at the sentence level of detection; the remaining three categories involve occurrence reporting at the sentence, sentence, and discourse levels, respectively. Categories (4) and (5), in particular, require much more extensive linguistic processing than the simple dictionary look-up of the first three.

The type of word-critiquing provided by (1) above would inform authors how many words in their composition came from "everyday" vocabularies, technical jargon lexicons, or — heaven forbid — special term glossaries. That of (2) would reflect the overall "tone" of the document in terms of such features as formal vs. informal, or colloquial/regional/slang, or neutral vs. evaluative, or friendly vs. unfriendly.[9] A report might indicate that the text, although quite informal overall, did contain several areas where excessively formal words were used. Category (3) involves the search for overused, cliche-type expressions which are more or less "fixed." For example, in the clause "We deem it advisable . . .," the key worn-out words are those in italics; one would hope to identify them no matter what variations of other words in and around them were made.[10] All three of these types of word critiquing are planned for the EPISTLE system.

While the first three types could be along quite well with relatively simple dictionaries and little further linguistic processing capability, the last two would require not only syntactic analysis of the text, but also very capable semantic processing. What is envisioned is the capability to provide the following types of diagnostics: "You probably meant to use the word <'infer'> here instead of <'imply'>" (class 4); "This word <WORD> simply doesn't make sense here" (class 4); "You are using the word <WORD> in two different senses in the same context" (class 5); or "Your use of the word <WORD> here implies one thing, but the use of the word <WORD> in the next sentence implies something very different" (class 5). Word usage critiquing of this type is well beyond the capability of any existing system.

Grammar critiquing

The notion of grammaticality is almost entirely restricted to a sentence unit. One doesn't have "ungrammatical" words — only words that are misspelled, oddly spelled, or, morphologically speaking, strangely formed. Similarly, paragraphs are text sections that may be called contentless, poorly organized, or highly redundant, but they are not deemed to be grammatically incorrect. The major class of sentence ungrammaticalities is that involving agreement — agreement in number, person, gender, and case. While English is one of the least inflected of languages, there are still a number of opportunities for grammatical disagreements.[11]

The detection of failures of agreements and other ungrammaticalities in a sentence requires that the sentence be "parsed" according to the grammatical rules of the language, resulting in a description of the syntactic structure of the sentence. This description must not only tell about what modifies what in the sentence, but also about the grammatical roles played by different parts — e.g., the subject, the main verb, the direct object. Parsing quality can be judged in two ways, both related to the adequacy of the syntactic descriptions. First, the modification structure can be judged as to its accuracy, the most important aspect being "were the words in the sentence assigned the correct parts of speech?" For example, the sentence (1)

(1) Time flies like an arrow

can be understood quite easily in the sense of (2)

(2) This airplane flies like a rocket

fairly easily in the sense of (3)

(3) The fruit flies do like an apple when they can get it

or with some difficulty in the non-obvious but still acceptable sense of (4)

(4) Please time the athletes in the manner of an Olympic judge

In the sense of (2) the word "time" is used as a noun; in the sense of (3) it is used as an adjective; and in the sense of (4) it is used as an imperative verb. For this example all three readings are possible, and the "correct" one is decidable only from other information. In these cases of true constituent ambiguity the structural descriptions provided by the parser will differ greatly; the follow-on processes which use these descriptions — critiquing, for example — will also vary widely. Thus, without complete understanding of the whole text, it is possible to make a reasonable misunderstanding of a single sentence.

Other problems involve the notion of "attachment." For example, in (5)

(5) I saw the man in the street

the prepositional phrase PP "in the street" either could specify which man or it could refer to the location where the "seeing" action took place. Therefore, the PP could be attached to either the noun phrase, NP, or the verb phrase, VP; in this example either could be argued. The question, though, is the general trend of attachments over a large number of parses: what is the safest course, and what are the various consequences? The IBM EPISTLE grammar approach, in the first version called here EPISTLE-1,[12] was to attach PPs to higher-level VPs rather than nearby NPs — unless other local information indicated otherwise (e.g., PPs with the preposition "of" were often attached to the immediately preceding NP). This approach, while not necessarily more correct in the long run than some other similar decision did have the consequence of "raising" these phrases in the parse tree, making them more easily examined. This kind of bias would lead to a certainly correct attachment of the PP for (6),

(6) I saw the man through the binoculars

a possibly correct attachment for (7)

(7) I saw the man with the binoculars

and a probably incorrect attachment for (8)

(8) I saw the man with the raincoat.

The second general way in which parser/grammars can be evaluated is by the accuracy with which they determine the grammatical function of various components of the sentence. Thus, an adequate parser in this sense would identify "John" as the "real" or so-called "deep" subject of (9) and (10) despite variations in the surface subject:

(9) John hit the ball

(10) The ball was hit by John

A functionally-accurate parser would also correctly identify the indirect objects in (11)-(14) as "him," "her," "whom," and "who" respectively:

(11) She gave him the book

(12) He gave the book to her

(13) To whom did she give the book?

(14) Who did he give the book to?

In addition to quality or accuracy, parsers can be judged also on their "robustness" in the face of ill-formed input text: does the parser wade through "noisy" errors and come up with something; or does it throw up its virtual hands and come to a screeching halt at the slightest difficulty? And, in particular, what does the parser do with sentence fragments? For these problems the original bottom-up EPISTLE-1 parser is also quite satisfactory, having several different strategies for handling — and surviving — problematic text, including unrecognized words, sentence fragments, sentence run-ons, ungrammaticalities, and the grammatically correct but troublesome instances of sentences with multiple legitimate (and illegitimate) parses (cf. Miller 1982a; Jensen & Heidorn 1982; Jensen et al. 1984). For the new top-down version of the EPISTLE parser now being implemented, called here EPISTLE-2,[13] there was initial concern that it would be quite difficult to use it for describing the partial and excess structures of sentence fragments and run-ons, respectively. However, it now appears that the same grammar can be used to accomplish these identifications with no re-parsing.[14]

Given that one has a parser, then — and there may be several available — the interesting question from the point of view of composition is the extent to which the parser can be used to supply grammaticality information about the text to the author. So far as we know, the IBM EPISTLE-1 prototype is the only one providing such function, and it provides for detection of 27 specific classes of errors, shown in Figure 3.

While the performance of the grammar can be enhanced, there is really not that much more that can be added to this set of grammatical critiquing targets. The practical problems remaining concern: (1) improvement in grammar performance,[15] (2) extension of the grammar-checking capability to new text domains involving different sublanguages and grammatical conventions, and (3) development of grammar components for other languages.

Style critiquing

In the critiquing of style we leave the firm safe shores of grammar absolute and push out to the seas of relativity: "This is perhaps awkward here"; "If your point is thus and so then it doesn't really come across"; "Do you really mean to say that?"; "Must you include this section?"; "I don't understand you here." Where there is more or less one single burning flame of grammatical correctness, there are hundreds of flickering candles of style, each attesting to the eternal truth of *"chacun a son gout"!*

Style is in the eye of the beholder; more precisely, style is the expression of an author's strategy for communicating information. For the texts of commerce, authors don't have to worry much about coming up with these strategies; their enterprise, their communication medium, and their field all act together to provide more or less explicit guides for style (as discussed in detail in the next section). Even more available is the capability to detect deviations from the "good" style of an author's place of business: either there's a "style handbook" lying around or else there's always an eager old hand who's quite willing to point out the numerous flaws of the stuff of the newcomer who has yet to learn the ropes about "how we write things around here."

Many people have initially denied the truth of this assertion, but, upon questioning, they all have had to concede that, yes, for every type of business or technical writing they did, they were either following the dictates of some crusty mental guide or else they could recall who or what was their source for good style when they had questions. For example, a business person will respond to a memorandum with a reply containing the same format, fields, and breadth of content; the quarterly sales report will be stylistically identical to that of the previous quarter; the notes to colleagues, the "instructions in my absence," the technical review, and the personnel appraisal all similarly will reflect thoroughly the accepted stylistics of that environment. Thus, if the first point made about style was variability, then the second point is quality control! Each enterprise sets the standards — as it sees fit — for controlling the form and conformity of each type of writing output. For better or worse, each contributor to the writing flow has to internalize these standards.

The contribution of computer software to this situation is the capability to provide each author a private forewarning of stylistic criticisms that might be delivered later from human agents. The computer critiques can be a much more agreeable situation, particularly when suggestions and explanations are

1. Subject-verb number disagreement ("The leaves falls")
2. Noun-premodifier number disagreement ("Each valves . . .")
3. Subject-complement number disagreement ("This is the books")
4. Subject-verb person disagreement ("I likes to eat")
5. Confusion of "it's" and "its" ("It's nose is dirty")
6. Wrong pronoun in predicate nominative ("The right person is her")
7. Wrong pronoun in object position ("Between you and I, it's a cinch")
8. Wrong pronoun as indirect object ("I will give they the book")
9. Confusion of "who's" and "whose" ("Who's side are you on?")
10. Confusion of possessive 's and plural s ("Few pet's are lonely")
11. Confusion of semicolon and comma ("Please buy coffee; milk; and eggs")
12. Conjoined verb number disagreement ("The sand both glitters and choke")
13. Incorrect double modals ("He should ought to go")
14. Existential "there" number disagreement ("There's five apples here")
15. Improper form of the infinitive ("He wants to won")
16. Improper form of the verb ("She broked my heart")
17. Missing obligatory commas ("He decided I think on the plan")
18. Missing obligatory hyphens ("I have a yet to be satisfied request")
19. Missing an obligatory question mark ("What is your name.")
20. Mismatching correlative conjunctions ("You are neither happy or sad")
21. Mismatch of the number of a noun phrase and its relative clause ("I saw the man who are here")
23. Sentence fragment or "non-sentence" ("Also my tires")
24. Unbalanced comma ("My manager, Mr. Hendrix told me about you")
25. Unmatched pronouns ("Him and I are going")
26. Use of wrong indefinite article ("I ate a apple and an pear")
27. Use of "of" for "have" ("I could of danced all night!")

Figure 3. Grammaticality critiques

given for each detected violation — as they are, with adjustable levels of detail, in the EPISTLE-1 system. For example, were a person to have typed the sentence (15)

(15) Between you and I, I am sure she will succeed.

the system, when asked to critique, would first underline the word "I" in red (the usual editing display is green text on a black background). If the person pressed the appropriate HELP button, a box within the text-editing screen would open up and a two-part (colored, reverse-video) message would appear saying: (1) Wrong pronoun in object position (prussian blue); (2) me (aqua marine; this strategy of (1) classifying the problem, and (2) suggesting something to do about it is followed for all three types of critiquing in the system). But then, a second press of the HELP button produces an extension of the message, a 20-60 word explanation of the tense classification; and a third press produces a detailed tutorial on the general topic involved, from half a page to three full pages in length. Thus, the author can not only locate each problem but can obtain replacement suggestions and increasingly detailed explanatory information in every case.

If such critiquing can help the peace of mind of the writer — and actually promote better writing skills — this is certainly of benefit to the enterprise, from the point of view of the satisfaction and competence of its employees. But, setting personnel concerns aside, such capability can provide the enterprise even more direct and tangible benefits by assuring that its written output is of constant high quality, according to its own dictates. Thus, at least proposals won't get dismissed because of poor grammar or awkward style, a newspaper will avoid being a laughing-stock because of a misspelling or tortured expression, a publisher's reputation for quality won't be smirched by missed "the . . . the"s (at the end of one line and the beginning of the next), and a manufacturer can have confidence that his assembly instructions are readable.

Where there are only a few dozen grammatical error possibilities at most, the number of stylistic targets — per individual enterprise — can easily be in the hundreds, and each major commercial area (e.g., education, legal, manufacturing, etc.) may have thousands, with only a few of these overlapping other areas. For example, the stylistics of legal writing are considerably different from all other types, but within this area, the specifics will vary greatly from one state of the union to the next, and even from one federal government agency to the next.

The consequence of this great diversity of styles is that a system purporting to provide maximally useful stylistic critiques must have been customized to the major commercial area, to the major subdivisions within that area, to the requirements of individual enterprises within these subdivisions, and even to departments within enterprises. We would further hold that every individual user must have some control over the stylistic targets — either to turn them on or off or to adjust their level of sensitivity.

EPISTLE-1 sentence-level style critiques

Many of the prototype's critiques are of the threshold type, in that the "strength" of the problem is computed on an increasing numerical scale, and this strength is compared to some threshold value; when the strength exceeds the threshold, the critique is voiced; when the threshold is not exceeded the system can quite sensibly keep its "mouth shut." These critiques, as well as the ones that are based on the direct occurrence of a pattern of present/absent symptoms, are far from being simplistic, in that the context is examined carefully before criticism — the same construction might be quite acceptable in one context but simply awful in another; the style critiques attempt such sophisticated discrimination.[16]

Almost the sole information source for the stylistic computations is the annotated parse tree which is output by the grammar component. This tree details the syntactic structure of the surface sentence, assigning grammatical roles to some elements (e.g., subject, direct object), providing information on all of the inflectional aspects and giving the semantic features associated with each text unit. The stylistic rules interrogate this tree, measuring it, counting aspects, comparing and contrasting, and develop the specific features needed for each critique target. Thus, the style component performs no additional general processing of the sentence or its parse-tree representation; measurements are obtained according to specific target descriptions directly from the tree.

The nature of the stylistic processing means that once the parse-tree has been obtained, the hard part is over; the rest derives from examination of that tree. The implication of this approach for obtaining new style targets is that it will be a relatively easy thing to accomplish — as long as the features that reveal the targets are derivable from the tree. Thus, we anticipate that it will be a relatively straight-forward matter for a specialist to implement new capabilities.

The source of most of the stylistic critique targets presently implemented is the general consensus that exists concerning "good" and "bad" style (see references in Miller 1982b). These will probably be of interest no matter what the details are of a particular enterprise or a user within that enterprise. However, each enterprise will have many specific "bad-style" targets, not included in the general set, and these will have to be custom developed. We estimate that perhaps 70% of an enterprise's special targets can be implemented to a reasonable degree by using information from a parse tree. The remainder (e.g., "Does each topic logically flow?", "Is it well organized?", "Is it informative and precise?") will be impossible to implement at the present time, because of two problems: (1) the critique requires complete semantic interpretation of the sentence, probably involving knowledge-based inferencing, and (2) the critique requires interpretation of large chunks of text, more than one sentence at a time — so-called discourse-level analyses. In Figure 4 are listed a sample of the 106 different style critiques presently implemented as EPISTLE-1 system function. Note that some punctuation errors are included; many such errors will rely on very extensive semantics for their detection and will not be achieved for some time.

1. Sentence too long (or paragraph, procedure step, or heading)
2. Sentence is not parallel
3. Too many nouns modifying the head noun
4. Too many consecutive prepositional phrases
5. Commas might be missing around non-restrictive modification
6. Preposition and participial are too far apart
7. Incomplete verb construction
8. Incorrect use of "like" as a subordinator
9. Incorrect use of comma as a complementizer
10. Possible excessive use of "and"
11. Incorrect comma between conjoined phrases
12. Possible excessive use of negation
13. Too many hedges/concessions
14. Overused phrase

Figure 4. A sample of EPISTLE-1 style critiques

The difficulty of writing these very general style critiques was compounded greatly by the requirement that the criticisms be reasonable over a wide variety of texts and authors, from college to the military, and from student to chief editor. We expect that it will be relatively easy to develop new critiques for a particular enterprise, especially one that is extremely "picky" and precise about its style criteria (such as a publisher).

Aggregate Style Critiquing

There are at present very few examples in EPISTLE-1 of this type of critique target, primarily because the others were more challenging and pressing to develop. We will include in EPISTLE-2, however, a number of aggregate critiques, summarized over larger text units of the document. For larger documents especially, we envision one form of the output of aggregate reports as being a kind of graph, with the horizontal axis marking off chapters, sections within each chapter, paragraphs within each section, etc. For one or more of these text units, then, frequency or percentage information on the following kinds of targets could be provided (the list is meant to be illustrative and is not exhaustive of the reasonable possibilities):

- Information Density
  - Words
  - Sentences
  - Facts[17]
  - References[18]
  - Specialized document entities[19]
- Word Features
  - Tonal aspects (formal, colloquial, etc.)
  - Use of rare or archaic words
  - Use of jargon or technical terms
  - Use of abbreviations or acronyms
- Sentence Features
  - Type (declarative et al.)
  - Voice (active vs. passive)
  - Complexity (simple, compound, et al.)
  - Readability

Discourse-level critiquing

The real aspects one wants to check in a document have to do with how effective it is in achieving its goals. Sentence-level style and diction are important, of course, but they do not begin to assess the quality of the overall document. At the discourse-level one certainly is interested in at least the following: Specific content, organization, coherence, relevance to topic, information-value, clarity. None of these can be assessed by computer software at the present time. And one pales at the thought of how much science and art have yet to be invented to make computer critiquing of this kind possible.

There is one possibility, however, for making some headway sooner, if authors will agree to some extra work. The idea is to use the tagging approach, discussed in the section on scribing and formatting, for authors to annotate their running text in terms of the rhetorical functions it is intended to achieve. The following is an imagined list of what a few of these tags might be:

:kprop. Key proposition or argument to be supported

:prem. Premise or assumption

:contx. Background or context of discussion

:tstruc refid=T1. Topic structure to be introduced, e.g., a set of 6 items to be discussed in turn; an element of this structure may refer back to this by using the reference identifier "T1."

:next ref=T1. Next item in the list spelled out by the topic structure identified as "T1."

Were an enriched and more carefully developed set of such tags developed, the analysis capability of, say, EPISTLE-2, would make it much more feasible to evaluate the free text enclosed by the tags in terms of the criteria listed above. It is not at all clear, however, whether authors could be so analytical as to identify the rhetorical functions of each text segment, nor whether they would tolerate having to use tags even if they knew. The point is that a compromise such as this does move forward the functional capability to do discourse-level critiquing — to perhaps just a few years away rather than a decade or more.

Format critiquing

A final type of critiquing concerns how the words are arranged on the paper (or computer screen), relative to how they should be arranged according to the writing-genre employed and the local customs. For example, it may be the policy of an enterprise for the title page of all technical reports to have a security classification as a running bottom margin item, report title centered ⅓ down from the top, the author's name followed by two blank lines and the address, etc.

The format critiquer should verify that all these formatting standards are met, first by identifying what type of document is being written, and then by checking the document against each of the standards. Since authors may not identify the document as being a memo, a report, a letter, or whatever, the system will have to have the capability to infer the genre — either from the embedded tags (if the source file is submitted for critiquing) or from the arrangement of blank space on the page (if the formatted file is submitted).

There's not much help in this area yet, but the requirements are clear, and implementation really depends more on standardization of document types and tags than upon the invention of new algorithms.

Commercially available systems

There are several commercially available systems which do provide some level of text-critiquing function. Perhaps the most well known of these is *Writer's Workbench,* which is marketed by Western Electric (cf. Cherry et al. 1983). The program essentially comprises four functional components: proofreading (spelling, punctuation, wordy phrases), style analysis (sentence length listing, distribution of sentences by type and complexity, sentence voice), information (English tutorials), and utilities (personalized dictionaries, new style standards). An extensive review of this product's function and performance is provided in Seybold Report (Seybold 1984); this report also reviews imitations of Workbench, the EPISTLE-1 system, and a number of other similar products.

The advantages of these products is that they are available now at rather reasonable costs, and they usually can be run on a stand-alone personal computer. The disadvantages may be that the function provided is necessarily limited by the size and development of the underlying language technologies — (1) the dictionary and (2) the parser and grammar; in most cases the dictionary contains only a few hundred function words, and the "parsing" is pretty much probabilistic guesses at parts-of-speech and identification of phrases — but nothing like a detailed description of the hierarchical syntactic structures.

In addition to the above limitations, there is no provision in existing commercial systems for accomplishing any semantic processing or knowledge-checking. Such capabilities not only will probably require large storage (a megabyte does not seem unreasonable), but also this type of processing can be quite involved and can occupy large amounts of CPU time. Extensions of function into these areas will most likely await the availability of much much faster and larger personal computers.

COMPUTERS AND THE REMAINING COMPOSITION STAGES

Whereas the previous discussions of scribing and critiquing have largely been focussed on actual software systems, the discussion of the remaining stages is largely speculative: systems to accomplish most of the described function have not yet been built. Nevertheless, the development of computer support for the first three stages would appear to be a somewhat easier task than, certainly, the development of large on-line dictionaries and parsers.

Computers and communication requirements (stage 1)

The key problem in this stage is determining what goals the text should meet; before these goals are firmly established there should be, ideally, no consideration at all of how the goals are actually going to be met. This phase actually occurs with familiar regularity in large text projects, particularly commercial ones. Thus, a book publisher will go through this phase in considering development of new titles, reference publishers will do so in considering the merits of issuing a new collegiate or learner's dictionary or junior encyclopedia, and technical journal editors will do so in thinking of a special issue on a particular topic. In all of these cases there will be careful consideration of the communication goals, the expected audience, the potential delivery media, the specific content to be covered, the various contexts to be related to, and all sorts of incidentals like who is it that has to be flattered by all this, and what are the possible details of schedule, pricing, resources, and follow-on needs.

But what of the lowly interoffice memo, a weekly paper in English Comp, or an outraged complaint letter? Is the communication requirements phase truly a useful aspect of these much more minor pieces? Since these all still have rich structure and pragmatics, the answer must be, yes. However, whereas the really big projects often have access to detailed guidelines for determining functional requirements, these are much less available for the genres in which the output is so much smaller and tidier. Ironically, it is these smaller genres, and not the larger, which have received some initial computerized support, albeit highly formulaic (cf. Schwartz and Bridwell 1984). One of these, WANDAH (Writing Aid and Author's Helper; cf. Von Blum and Cohen 1983), provides this component in the context of several other aids, including a text editor. An advantage of such programs is, of course, that one can begin to see the potential for true support of composition throughout the whole process. The difficulty is that these programs have no way of determining whether students' responses to their prompts are in any way reasonable — or even relevant.

Ideally, computer support of this phase would involve three aspects. First, the computer would present the would-be author with a set of "considerata" to reflect upon and respond to. These would be derived from, and ordered in accordance with, a detailed psychological model of this very early stage of composition. Second, the computer would assess the quality (detail, clarity, relevance, etc.) of the author's response to the "considerata"; were the quality insufficient, the computer could provide feedback (e.g., prompt, chide, fret, or bully). Third, and by far the most difficult, the computer could assess the product of later stages of composition in terms of how well they implemented the requirements; detection of incompleteness or contradiction could trigger bringing the author back to consider again the functional goals set forth way back when, to resolve the discrepancies. None of these three things is likely to happen for a while, since there's nowhere near the necessary and proven cognitive theory to support such work. Even if the theory were available, there's hardly a clue how to achieve implementations of this kind.

Computers and high-level planning (stage 2)

In this stage the author leaves consideration of what should be achieved and ponders the how of it. For example, suppose the communication requirement is mainly to get back at some agent of commerce who refused to replace or credit a defective piece of merchandise. It's not at all certain that the communication action must involve a letter to someone in that enterprise; perhaps, instead, an ad should be taken in the *Times* decrying the action; or perhaps a telegram should be sent to a friendly and influential senator. The decision to be taken at this stage is to develop a broad general approach to the problem, after considering a broad set of different approach concepts. If everything that one has thought about in this stage is but a slight shade off from one another, then likely one has not pondered long—nor "High"—enough.

If there is doubt about being able to support the prior stage with computers, there is at least as much uncertainty about it for this stage. What is really at stake here is the act of developing an effective overall plan of action from a set of goals. What is needed is a pragmatic handbook which gives details of procedure, and evidence of success, for the cornucopia of possible communication plans and approaches matched to goals. Were this guide available, it could most certainly be converted for on-line use. Less certain, of course, is the extent to which authors' communication requirements statements can be evaluated against subsequent plans to accomplish the stated goals.

Barring such a handbook, there's not very much that can be computer-implemented at this point except for very general open-loop guides. There are in fact several software packages of this sort which are inteded to help the writer develop a working plan which moves easily into the actual scribing task (see Seybold 1984; Schwartz and Bridwell 1984; Nancarrow, Ross, and Bridwell 1982; Dalgish 1984). Two limitations with these programs are that they are (1) very general and undifferentiated with respect to specific writing topics, and (2) they have no way of evaluating the students' input.

Computers and detailed planning (stage 3)

At this point the author has narrowed the possible approaches to a single major one and now must ponder the details: e.g., "A letter, surely, but how phrased? . . . and could I be sued for libel?" Since computer support of design has been adequately reviewed in the previous section, and since no one else seems to differentiate between high and detailed design, what is given here is a proposal for providing rather strong and closed-loop detailed design support in a manner which moves quite easily into the following stage of scribing (and even critiquing).

In the absence of data to the contrary, it is useful to presume that one of the key decisions of the high-level design was to decide upon the major type of communication genre to be employed. That is, you know roughly whether you have to write a report, a letter, a legal document, a journalistic article, or a fictional story. The essence of the present proposal is to have the computer provide you with a way to narrow down the broad set of choices that still remain until you have chosen a quite specific document genre, and then to be able to tell you all the details of the pragmatics of this document so that you can start adapting it to your particular needs.

The first key element is the detailed classification of genres within an area, organized in such a way that writers can find their way to a specific sub-genre which will meet their needs. Considerable thought was given to an approach towards classifying major genre areas to minimize overlap and to reflect human pragmatics strongly, and the choice was to partition major genre branches according to the major social institutions: law, medicine, education, military, the press, government, commerce, etc. Two of these branches, law and the press, were chosen to be developed, to get some feeling for the proposal.[20]

The first task was to develop a workable taxonomy of writing genres within each field, organized from the viewpoint of the demands on an author to write within a particular genre. About 250 sub-genres were decided upon for both law and journalism, and the first few levels for law are shown in Figure 5 (Halpern 1984a).

This taxonomy (and the one for Journalism as well) was developed and tested iteratively by repeated examination of (1) diverse textbooks and (2) many actual examples. The development of the finest genre distinctions, the sorting of these in the hierarchical tree, and the naming of the classification nodes themselves, all continued until new input was easily identified and caused no problem in classification. These sub-genres were also cross-classified in terms of their values for a number of independent dimensions relating to their rhetorical quality. A variety of different dimensions were considered, and those that were retained (1) had some face validity with respect to rhetoric and communication and (2) showed a reasonable variation in values across the several hundred sub-genres. The dimensions settled on for Law are shown in Figure 6 (Halpern and Miller 1984, 13-14).

With the taxonomy and the dimensional classification systems then, the first part of the proposal was accomplished: Authors, knowing that the communication objective required writing a legal (or journalistic) document, could find their way to an extremely specific sub-genre which could accomplish their purposes. The second step is to provide the computer support to assist the authors in understanding all aspects of this genre, and to provide the framework for moving easily into scribing from this stage for elements of the document, and then back again into detailed planning.

The means for supporting the second step was to permit an author to choose a sub-genre, be put into an editor showing the high-level skeleton of that genre, and then be able to inquire about three aspects of any part of that skeleton: (1) the sub-classification, (2) the writing conventions, and (3) the rhetorical functions. Figure 7 shows the top level structure of a dwelling lease, in which an author has asked for elaboration of the property aspect, and then, within that the facilities included (Halpern 1984b).

By positioning the cursor on a line and pressing the appropriate key, an author can accomplish the three functions desired, that of expanding the category into its next levels shown in the figure. To determine the rhetorical importance of asking about the garage, for example, the author would position the cursor on that line, and press the *Rhetorical Purpose?* key; in this example, there would be a short response pointing out that garages are often contracted for separately and are often a basis for dispute if not specified clearly, etc. Pressing the *Writing Convention?* key might produce information that the key aspects of the garage to be covered are (1) its deviation for "normal" size (perhaps can accommodate only compacts), (2) number of bays, (3) whether it opens on a street or in a private area, and (4) whether there is room for storage. If this aspect called for special formatting, use of capitalization, or other nontextual aspect, that would be presented also.

1. Legal argument / scholarship
   a. Motion / petition
   b. Brief / correspondence / opinion
2. Records
   a. Affidavit / bill of evidence
   b. Deposition / record / minutes
   c. . . .
3. Pleading, orders
   a. Summons / warrant / citation
   b. Decree / order / injunction
   c. . . .
4. Legal transactions
   a. Free form (contract, offer, convenant, will, . . .)
   b. Fixed form (bond, stock, offering, order, . . .)
   c. Non-common law
      1) Legislative (bill / statute / ordinance)
      2) Constitutional (amendment / referendum)
      3) Admin / executive (treaty / regulation)

Figure 5. Taxonomy of legal writing genres

1. Density of promissory statements
2. Density of opinions
3. Extent to which authority is cited
4. Relative mix of sentence types (declarative et al.)
5. Density of boilerplate language
6. Extent to which the form is prescribed
7. Density of terms of art
8. Density of abstraction
9. Characteristic expressions and terms of art

Figure 6. Dimensions of legal style

1. Document title
2. Date of this document
3. Parties
4. Description of property
   a. Address
   b. Facilities included
      1) Number of rooms
      2) Description of each room
      3) Whether garage is included
   c. Period of lease
6. . . . (and 16 additional sections)

Figure 7. Structure of a dwelling lease

The usefulness of this approach from the detailed design point of view is that authors can be helped to rapidly identify and learn about the sub-genre that will meet their communication goals. In addition, however, it is easy to see how an author can keep right on going into the scribing phase: having seen the lowest category level (e.g., "whether garage is included"), it would be quite easy for the author to simply generate appropriate text for that section, according to the rhetorical function and writing convention information supplied. Then, having written something, the critiquing phase can be called upon to use the detailed tags, and all the other associated information, to assess whether what was written was relevant, coherent, effective, etc. — all those impossible discourse-level aspects discussed previously.

While this approach has yet to be implemented and data on its utility obtained, still it does provide a constant high level of support for the composition process through most of its important phases. Whether this particular concept is the best or not is of little consequence relative to the possibility raised here of more effective computer support — and once again through the use of identification tags.

AFTERWORD

After such a trek through the composition process, it is good to reach a vantage point and consider longer views. We shall take two, one looking at other audiences, and the other looking at the roles educators can play in this, their new computer future.

Specialized audiences

The intended audience for the various computer aids was never really discussed in the foregoing, although technical writers were surely a focus. But nothing said so far really precludes the traditional educational audiences that are briefly considered here.

College. A phenomenal upsurge of interest in composition, particularly computer-based, has occurred in the last few years. New sub-departments — even departments — of rhetoric are formed (or dusted off), linguists are added or adjoined to the English faculties, and there's tremendous interest in the other Pascal and things like operating systems, non-human editors, RAMs, and bytes. Of all of the educational areas, it is in the colleges and universities that there is the greatest interest in computers and composition, and perhaps the greatest capability to pay and staff such activities. Our experience in examining college compositions and attempting to run them through our EPISTLE-1 system suggests that such an interest may be well-motivated: there is certainly more, in the area of style and diction, for these students to learn than might have been expected. Clearly, there is considerable demand for particularly the critiquing capability, but considerable modifications will have to be made to existing technology to survive the hardships that some of the less well-trained students will place on the system.[21]

9-12. High school students may well be the most amenable to working with the guiding function concept described in the section on computers and detailed planning.

K-8. The underlying language technologies, if properly interfaced, could well provide exceptional opportunities for training the language newcomers in the basic skills of this craft. For example, were a large dictionary to be accessible in a variety of interesting ways, children could, greatly through self-discovery, learn that words are composed of parts (some fairly standard), that words which look different might sound alike (and vice versa), and that the meaning of words provides a rich set of connections among them (some positive, some negative, some generalizing, some specializing). Similarly, the hidden non-linear structure of a linear sentence could be revealed, showing parts of speech, phrases, and grammatical function, all changing as this or that part is changed by the student.

Remedial. The computer offers special promise for those needing extra assistance because their needs can so often be precisely determined and prepared for. A side benefit, but one which may well greatly increase the will to learn, is that the help, discovery, and mastery can occur in a private dialogue between student and terminal.

ESL. English is a second language for millions of people in this country, only a few of which are adequately enrolled in the school systems. The rest must make do with inadequate night training, books, and the help of friends, and, of course, their learning is slow, limiting their potential enculturation, assimilation, and opportunity for better employment. It may be that a number of the errors a non-English speaker makes can be predicted from the natural tongue (the remainder are developmental and also well known), and therefore specialized critiques can be developed in a rather straightforward manner. The potential benefits to this population, then, are enormous. The other aspect of ESL is that of foreigners, quite happily living in their own country, but with a need to write in English. Since poor English can be such a stigma, causing potential buyers to look elsewhere, the commercial possibilities associated with this opportunity are very great.

Role of the educators

The last item of concern is the potential reaction of those who must ultimately midwife the computer technology of composition. When faced with the support capability discussed here, educators and other administrators can adopt a variety of positions, fairly well covered by the following four: (1) they can ignore the whole movement; (2) they can oppose it; (3) they can evaluate it; and (4) they can shape it.

The first position is unlikely to be a quiet one, what with students clamoring for word-processing and more, the school administration pushing for greater "productivity," and parents going on about higher quality education. The second position of actively opposing these trends, while certain to make the front pages for a while, is much more dangerous if, as seems very likely, this capability is not going to go away, but grow rapidly stronger. Determined opposition may very likely leave one arguing at the backs of the retreating educational mass. The third position of testing the technology is certainly very constructive and absolutely necessary to some degree, if only to keep the salesmen's claims honest. The problems are that this is (1) awfully costly in terms of money, time, and personnel, and (2) it may well be quite premature, since the field is changing so radically. Only the very flexible and fleet of well-heeled foot can really succeed in this post.

The last position is one in which everyone can participate, but is the least popular at this time. Nevertheless, the development of effective computer-based writing tools requires the melding of the two disciplines: the computer-linguist and the educator. Neither can hope to do a truly reasonable job without the other, but it is the computer developers that won't let a little thing like that stop them! What the educator can do, then, is to understand better the nature and limitations of the technology and then help the developer bend his wares towards those areas in which the most gain can be made.

One rather painless way an educator can provide useful stimulus is to think practically about what kind of specific writing function is needed most. These ideas can then serve as functional requirements to the computer linguist to design and build more effective and useful systems — sooner. To put it more succinctly: *Educators: Get involved, Now!*

1. A more complete model usually has a "problem formulation" stage preceding the first stage of functional requirements, an expansion of the fifth stage of testing into three stages ("unit test," "system integration," and "system test"), and an addition of another stage somewhere after installation—"Maintenance").

2. Saying that such-and-such activity occurs in this or that stage is not an observational generalization. However there is substantial evidence, not easily dismissed, which suggests that design processes — whether of text or of programs — tend to have more desirable properties when, in fact, the activities and the stages are matched as described (e.g., Malhotra et al. 1980; Carroll et al. 1980). In this sense, then, the model is a normative rather than a descriptive one.

3. In earlier polyglot times an additional decision would also be approached — that of the language to be chosen (e.g., French for matters of philosophy, German for technical subjects, etc.).

4. In theory there is no reason why the last word just entered on a page is not a revision rather than initial scribing: a word could have been chosen, passed to the motor apparatus for output, interrupted there by a higher process which was reviewing the text in parallel, and a new word substituted. An observer would have no way of knowing whether that word was the pristine "next" in an unreflected stream or the *n*th choice after *n-1* rejects.

5. In contrast, for programming, this stage is an important area for the development of specialized tools to help the program-modifiers learn about the foreign programs they have been asked to alter — since the modifier of a program is seldom the original author (cf. Miller 1978).

6. Consider the sentence "The plane stops in Dallas/Ft. Worth". Many word-processor users might well be suprised to learn that the fifth "word" (token) processed by the computer is Dallas/Ft.

7. In the IBM system, headings are indicated by the prefix letter "h", followed by an integer from 0 to 9; as the integer increases so does the fineness of subdivison. Thus, "h0" is for a major part of a document, "h1" for a chapter, "h2" for a major chapter subdivision, etc. The sequence of section numbers given in the text ("1.0 . . . 2.1211") could be produced by the following sequence of tags in the text: :h1. :h1. :h2. :h3. :h3. :h4. :h5.. The text associated with each level tag could be varied in font type, size, and boldness, according with the local formatter interpretation programs.

8. A very simple example is that of the word "past," for which the dictionary returns the following string of part of speech features: (past(ADJ) past(Noun Sing) past(Prep)). For other words, especially verbs, many other features and attributes are reported in addition to part-of-speech. The dictionary access method and structure are the responsibility of Roy Byrd, with assistance from Mary Neff, Martin Chodorow, and Barbara Kipfer (and, previously, from Barbara Snitzer and Lori Alperin).

9. Tone here refers to the "connotations of interpersonal attitudes" and is primarily a word-level phenomenon. While most dictionaries do not report these connotations (as a function of particular sense or context), there is certainly no reason that this type of information could not be added to on-line dictionaries.

10. Thus, one would want to flag all of the following for the same hackneyed aspect: "This council deems this action advisable," "Do you now deem that an advisable course?", "We have deemed this, and all other actions taken by the previous administration, as reprehensible and not advisable pursuant to the . . . " On the other hand one would probably not want to flag the following sentences in the same way: "We deem him honorable, whether or not his actions were advisable in the circumstances . . .," and "Most deemed it wise to go slow. We surely all believe in the most advisable courses of action." Detection of frazzle phrases under such variable circumstances requires the aid of a grammar analyzer in addition to the dictionary.

11. Grammar-checking in Russian, Latin, and Hungarian, for example, would be a very much richer field, given the large number of noun cases, declensions, and conjugations of these languages: German grammar-checking, while less complicated, still has four cases and three genders, thereby offering many opportunities for ungrammaticality — more than, say, Spanish.

12. This original EPISTLE grammar employed the NLP rule language and bottom-up parsing approach of George Heidorn (cf. Heidorn 1972). It has been implemented in LISP and, recently, in a variant of PL/I, and it could accurately be called a "syntactic grammar" in that almost no semantic information was used in obtaining reasonable parses for a wide coverage of English—something of a surprise to many of our colleagues. EPISTLE-1 was the first to demonstrate the feasibility of accomplishing quite sophisticated syntactic analysis and text-critiquing; its grammar was primarily the work of Karen Jensen. The parses of EPISTLE-1, while adequate for a variety of text critiques, were still "appropriate." For example, the structural descriptions were not always right, some grammatical roles were not identified (e.g., some indirect objects), and the syntactic association between discontinuous but related elements might not be recognized (e.g., in [14], the relation between "who" and "to"). In addition, while semantic features could

be added to the grammar rules, there were no general mechanisms to accomplish such things as sense selection, semantic-case assignments, knowledge-checking, quantification, and overall scoping.

13. The new EPISTLE-2 PROLOG version is designed to produce an "accurate" first parse ("rather than "approximate") to support a wider range of applications requiring more semantic interpretation — including document indexing, and mechanized translation. Semantic features and type hierarchies, grammatic slot-frames, and knowledge-checking are all employed to achieve correct sense-selection and case assignments integral to obtaining the "accurate" parse. A logical form can also be produced which represents the meaning of the sentence in predicate expression form. The semantic formalisms and PROLOG implementation are presently those of Michael McCord (cf. McCord, 1982, 1984, 1985), who also provided an extensive startup PROLOG grammar of English. With this beginning, Alexa McCray is developing a theoretically motivated system grammar, adapting McCord's interleaved syntactic/semantic approach to our requirements, ultimately to include handling of discourse-level phenomena.

14. Martin Chodorow is thanked for his insight into PROLOG and the grammar, and it appears that he has identified an elegant solution for the fragment and run-on problems.

15. The EPISTLE-1 grammar and critiquing function has been tested repeatedly on a set of 2K sentences and over a variety of texts, including business correspondence, student themes, and military manuals. Many sentences are given a reasonable parse, and most of the errors are detected — those that naturally occurred and those that were introduced specifically to test performance. On the other hand, a number of sentences received an incorrect parse, and the false-alarm rate (saying there's a grammar problem when the text is actually correct) is quite a bit higher than we wanted. Our plans for the EPISTLE-2 prototype are to create a testbed of 4-8K sentences (from perhaps a dozen genre types, mainly commercial), to develop automatic means for characterizing the features of a parse (comparing the latest parse of a sentence to that from the previous testing session, and to the "ideal" parse if available, and automatically classifying the nature of differences; this is the work of Mary Neff), and to classify the parse overall as perfect, acceptable for critiquing, or unacceptable, summary statistics will be provided for correctness (of the first, desired, parse), total number of parses, CPU time, and storage. Our goals are to develop an experimental system which provides at least acceptable parses for 98% of the sentences, detects 99% of the grammatical errors, has a false alarm error rate of less than 1%, and runs quickly and comfortably on an IBM PC-AT.

16. Yael Ravin has developed and implemented the style function.

17. A "fact" is either a proper name or a certain type of quantified noun phrase. Proper names are names of people, places, or things, and the latter two categories can get quite complicated. Quantified noun phrases include money amounts, physical dimensions, addresses, dates, phone numbers, etc.

18. References are explicit stylized pointers to other sections of the document and are of five types: (1) pointers to items in the reference bibliography (e.g., "McCord 1984"), (2) pointers to footnotes, (3) pointers to entities embedded in the text (e.g., formulae and examples), (4) pointers to non-textual entities (e.g., figures, tables, photos) and (5) pointers to structured text units (e.g., "see Appendix 1.2," "cf. Chapter 3;" a page-number reference may be considered such a pointer).

19. These are items other than straight text, to which reference is usually made in the text, such as footnotes, tables, formulae, etc. They also include other items to which no reference may be made but which are visually — and informationally — an important aspect of the document (e.g., company logo, printed telephone number, text border, dividing line between text and footnote, etc.) Repeating page elements, such as a running title or a security classification, are not considered as instances of this type.

20. In the summer of 1984 Joseph Janes and David Halpern of the School of Information Studies, Syracuse University, worked with the author to flesh out the detailed proposal for their respective fields of journalism and law. All details provided here are derived from that work program. The legal work was chosen for illustration (Halpern 1984a, b; Halpern and Miller 1984), but the journalistic work (Janes and Miller 1984) was as comprehensive and complete.

21. We find that students run words together, break words at other than hyphen points, leave out spaces between an end-of-sentence period and the first word of the next, capitalize in strange ways, omit obligatory punctuation, repeat words unintentionally, accidentally shift their hands on the keyboard while touch typing, introduce spurious characters (particularly when they hit a repeating key by mistake), skip lines randomly and do various other system-wrenching things, not even mentioning some of their peculiarities of syntactic structure.

REFERENCES

Byrd, R. J. "Word formation in natural language processing systems." *Proceedings of the 8th International Joint Conference on Artificial Intelligence,* 1983, 704-706 (Karlsruhe, Germany).

Byrd, R. J. "The ultimate dictionary users' guide," IBM Research Internal Report, 1984.

Byrd, R. J., and M. C. McCord. "The lexical base for semantic interpretation in a prolog parser." Presented at the CUNY *Workshop on the Lexicon, Parsing, and Semantic Interpretation,* January 18, 1985, New York City.

Card, S. K., T. P. Moran, and A. Newell. "The manuscript editing task: A routine cognitive skill." XEROX Palo Alto Research Center Report No. P76-00082, 1976.

Carroll, J., J. Thomas, L. Miller, and H. Friedman. "Aspects of the solution structure in design problem solving." *American Journal of Psychology,* 1980, *93,* 269-284.

Cherry, L. L., M. L. Fox, L. T. Frase, P. S. Gingrich, S. A. Keenan, and N. H. Macdonald. "Computer aids for text analysis." *Bell Laboratories Record,* May/June, 1983, 10-16. 269-284.

Dalgish, G. M. "Microcomputers and teaching English as a second language." *City University of New York Research Monograph Series,* Report No. 7, September, 1984.

Durham, I., D. A. Lamb, and J. B. Saxe. "Spelling correction in user interfaces." *Communications of the ACM,* 1983, *26*, 764-773.

Embley, D. W., and G. Nagy. "Behavioral aspects of text editors." *Computing Surveys,* 1981, *13,* 33-70.

Flower, L. *Problem-solving Strategies for Writing.* New York: Harcourt Brace Jovanovich, 1981.

Flower, L. and J. R. Hayes. "The cognition of discovery: Defining a rhetorical problem." *College Composition and Communication,* 1980, *31,* 21-32.

Halpern, D. J. Unpublished taxonomy of legal writing genres, 1984a.

Halpern, D. J. Unpublished description of dwelling leases, 1984b.

Halpern D. J., and L. A. Miller. "Automating legal writing." Draft of a report on a classification of legal writing and the implementation of computer support for three selected sub-genres, August 27, 1984.

Heidorn, G. E., Jensen, K., Miller, L. A., Byrd, R. J., and Chodorow, M. S. "The EPISTLE text-critiquing system." *IBM Systems Journal,* 1982, *21,* 305-326.

IBM "Using the professional office system." IBM Publications Manual, Document Number SH20-5604-0, Chapter 12 ("Proof reading documents"), June, 1983.
IBM "Document composition facility: generalized markup language, starter set reference." IBM Publications Manual, Document Number SH20-9187-2, April, 1984.
Janes, J. W., and L. A. Miller. "Towards automated guiding and critiquing of journalistic writing." Draft of a report on work accomplished in classifying legal writing and implementing computer support of three sub-genres, August 9, 1984.
Jensen, K., and G. E. Heidorn. "The fitted parse: 100% parsing capability in a syntactic grammar of English." *IBM Research Report,* RC 9729, 1982.
Jensen, K., G. E. Heidorn, L. A. Miller, and Y. Ravin. "Parse fitting and prose fixing: Getting a hold on ill-formedness." *IBM Research Report,* RC 10457, 1984.
Leerburger, B. A. "IBM spelling checker." *Think,* November/December 1981, 25-30.
Malhotra, A., Thomas, J., Carroll, J., and Miller, L. "Cognitive processes in design." *International Journal of Man-machine Studies,* 1980, *12,* 119-140.
Mathews, J. C., and D. Stevenson. *Designing Technical Reports: Writing for Audiences in Organizations.* Indianapolis: Bobbs-Merrill, 1976.
McCord, M. C. "Using slots and modifiers in logic grammars for natural language." *Artificial Intelligence, 18,* 327-367, 1982.
McCord, M. C. "Focalizers, the scoping problem, and semantic interpretation rules in logic grammars." *Logic Programming and its Applications,* D. Warren and M. van Caneghem, Eds., 1985.
McCord, M. C. "Semantic interpretation for the EPISTLE system." *Proc. Second International Logic Programming Conference,* 65-76, Uppsala, Sweden, 1984.
McCord, M. C. "Modular logic grammars." *IBM Research Report,* RC10936, 1985.
Miller, L. A. "Behavioral studies of the programming process." *IBM Research Report,* No. 7367, 1978.
Miller, L. A. "Project EPISTLE: A system for the automatic analysis of business correspondence." *Proceedings of The First Annual National Conference on Artificial Intelligence,* Stanford University, August, 1980, 280-282.
Miller, L. A. "Natural language text are not necessarily grammatical and unambiguous. Or even complete." *Proceedings of the Association for Computational Linguistics,* Toronto (June 18), 1982a. (Previously appeared as IBM Research Report RC 9441).
Miller, L. A. "How to write: a barely annotated bibliography." *IBM Research Report* RC 9687, November, 1982b.
Miller, L. A., and J. C. Thomas, Jr. "Behavioral issues in the use of interactive systems." *International Journal of Man-machine Studies,* 1977, *9,* 509-536.
Miller, L. A., G. E. Heidorn, and K. Jensen. "Text-critiquing with the EPISTLE system: An author's aid to better syntax." *AFIPS Proceedings of the National Computer Conference,* Chicago, May 4-7, 1981, 649-655.
Misek-Falkoff, L. D. "Computing, text markup, and 'speech acts': From outline to published book with authorial ease." Paper presented at the *Sixth International Conference on Computers and the Humanities,* North Carolina State University, Raleigh, June 8, 1983.
Morris, C. *Signs, Language, and Behavior.* New York: Prentice Hall, Inc., 1946.
Nancarrow, P. R., D. Ross, and L. Bridwell. "Word processors and the writing process: An annotated bibliography". Privately made available, English Department, University of Minnesota, 1982.
Orr, K. T. *Structured Systems Development.* New York: Yourdon Press, 1977.
Pollock, J. J., and A. Zamor. "Automatic spelling correction in scientific and scholarly text." *Communication of the ACM,* 1984, *27,* 358-368.

Roberts, T. L. "Evaluation of computer text editors."XEROX Palo Alto Research Center Report No. SSL 79-9, 1980.
Schwartz, H. J., and L. S. Bridwell. "A selected bibliography on computers in composition." *College Composition and Communication,* 1984, *35,* 71-76.
Seybold. "Computer aids for authors and editors: A natural extension of word processing and typesetting?" *The Seybold Report on Publishing Systems,* 1984, *13(10).*
Sekuler, R. "From quill to computer." *Psychology Today,* February 1985, 36-42.
Yourdon, E., and L. L. Constantine, *Structured Design: Fundamentals of a Discipline of Computer Program and Systems Design.* Englewood Cliffs, NJ: Prentice-Hall, Inc. 1979.

The *UNIX*™ Writer's Workbench Software: Philosophy

Lawrence T. Frase
AT&T, Warren, New Jersey

Technology has dramatically increased the number and complexity of written documents. The Bell System spends over $100 million annually on technical documents, with much effort devoted to review and revision. The UNIX™ *Writer's Workbench programs assist documentation by automating copy editing and proofreading tasks. These programs deliver detailed measures and comments about text readability, punctuation, word use, abstractness, and other features. Authors use the system to evaluate draft documents, and many feel the programs improve their writing skills. The programs have been used for quality control and text research, and in writing courses. This paper discusses principles used in system development and suggests possible uses for automated language analysis aids.*

I. INTRODUCTION

Massive growth in the number of documents and in the sophistication needed to understand technical language has created challenges for many organizations, outside and within the Bell System. The Naval Air Systems Command, for instance, supplies technical manuals for 135 aircraft. There are over 25,000 manuals, totaling 3,000,000 pages. In 1950, the manuals for one aircraft contained fewer than 2,000 pages; today, the manuals for one aircraft contain nearly 300,000 pages.[1] In the Bell System, there are over 100 general categories of documents. Just one of these categories, the Bell System Practices, contains 400 broad document categories, with 35,000 active titles. Current issues are routinely distributed to 45,000 locations. But number of pages is not the only concern. In some states, if documents are relevant to the physical or financial well-being of individuals, laws dictate how difficult the language of a text may be. Recent "plain language" legislation, passed by six states, allows courts to rewrite texts that are unclear.[2]

The telecommunications industry, along with others, is faced with two substantial problems: (1) how to manage a large number of documents, and (2) how to make them understandable. Review and evaluation of technical documents is costly, time-consuming, subjective, and often tedious. Mechanized aids for reviewing documents, although not relieving humans of responsibility for document quality, can simplify and speed the documentation process. Efforts elsewhere in Bell Laboratories address the problem of managing a

The Bell System Technical Journal, Vol. 62, No. 6, July–August 1983, 1883–1890. Reprinted with permission from the AT&T Technical Journal.

large number of documents.[3,4] The *UNIX** Writer's Workbench software was designed to help improve quality.

Until recently, document technology has consisted of two parallel, but separate, lines of work. The first consists of experimental research on what makes text easy or difficult to understand. This work, reflected in psychological and linguistic research, includes studies of the effects of text format, wording, and graphic displays.[5] The second line of work includes more intuitive or prescriptive bases for writing and text design. This involves the expertise of text and graphics designers,[6] such as technical editors and layout specialists, as well as rhetorical traditions that specify writing standards.[7] Both knowledge bases, one derived from experimental research and the other from nonexperimental sources, provide useful guidelines for effective text design.

Unfortunately, text researchers and designers have not always worked closely together. Psychological theories of what makes a document difficult to understand often seem oversimplified to designers, and designers' intuitions and prescriptions often seem vague to researchers. Designers have been concerned with creating good documents, while researchers have been trying to understand problems of text comprehension.

The use of computers for language analysis is beginning to stimulate integration of technical knowledge from research and design areas. McMahon, Cherry, and Morris,[8] for instance, have shown how language analysis, such as vocabulary analysis or assigning parts of speech to words, can be done easily and accurately by computer. Humans find those tasks difficult and tedious. Models of decision processes used by experts have been incorporated into programs that make design recommendations, such as the most suitable column width to use for a particular text.[9] Knowledge-based expert systems,[10] developed by researchers in artificial intelligence, provide additional ways of expressing, in computer-readable language, expertise derived from different disciplines. Development of the Writer's Workbench system has required, and continues to encourage, this contact between different disciplines. The programs recognize the validity of principles derived from research and the validity of principles derived from intuition and practical experience. Hence, they help bring research and application closer together.

The Writer's Workbench system provides a simple set of commands that deliver many assessments needed in documentation work. These include editorial comments on punctuation, word use, spelling, and text abstractness, along with analyses of grammatical parts of speech and calculations of overall text readability (expressed as reading grade levels). The programs combine work done by the Computing Science Research Center and the Human Performance Engineering Department of Bell Laboratories.

Using routines already developed for *UNIX* software, such as spelling and parts of speech analysis programs, developers were able to build programs that expanded and went beyond existing facilities. For portability, the wide

*Trademark of Bell Laboratories.

distribution of *UNIX* operating systems in Bell Laboratories was an advantage. An early version of the programs was tried on several *UNIX* systems for four months before the first general release within Bell Laboratories. During those months, program efficiency, outputs, and options were modified in response to user feedback. Although current versions of the programs incorporate further enhancements and pruning of esoteric programs, the system architecture has remained the same.

Text to be analyzed must be stored in the computer. Some authors prefer to compose text on line; others have it entered by clerical staff. Text should contain standard formatting commands, for instance, commands that define headings and other special elements. To analyze a text, the user logs in to the system and issues a command name followed by the name of the file that contains the text. The program then executes, prints analyses, and perhaps suggests changes to improve the text. Currently, the programs are appropriate for proofreading and stylistic analyses of expository and descriptive prose, but not for procedural documents consisting entirely of lists or abbreviated step-by-step directions.

The Writer's Workbench system is limited by the text features that the programs can recognize, and also by our understanding of the relation of text features to reading difficulty. For instance, one program recognizes when the first word in a sentence is not capitalized; it is more difficult to develop a program that can recognize when important content is missing. Currently, the programs do not use linguistic parsing algorithms; they cannot "understand" the subject and object of a sentence, for example, as humans would. Nevertheless, the richness and variety of feedback delivered by the programs are different enough from commonly encountered word and text processing systems that we find it useful to refer to these expanded capabilities as "language processing" functions. Hence, we have used this terminology here and in the following two papers, perhaps as much to remind ourselves of what remains to be done as to characterize the current programs.

II. DESIGN PRINCIPLES

Six specific principles, reviewed below, guided Writer's Workbench program development. The principles provide a context for the detailed program descriptions in the paper that follows this.[11] To summarize, the principles assume that language analysis programs should be rational, diverse, evaluative, modifiable, specific, and informative.

2.1 Expert knowledge base

The Writer's Workbench programs are *rational* in the sense that they are based on research or expert consensus that justifies the relevance of program information for text comprehension or use. Some program outputs are based on psychological research, showing, for instance, that passive sentences are more difficult to understand than active sentences.[11] Other programs are based on advice from writing and style experts.[7,12] For instance, standard punctuation usage is evaluated by some programs. Knowledge about the features of good writing is scattered among disciplines; hence, no single discipline is entirely adequate for formulating language analysis programs. Currently, we have drawn from rhetorical and psychological literature to provide text assessments that people commonly use in document evaluation. To do more, contributions from artificial intelligence and related fields are required. In addition, since not all features of document design can be clearly supported by research or expert opinion,[13] programs contain messages directing users to information sources that explain certain evaluations.

2.2 Program diversity

Written communication is so complex that single measures, such as an overall index of text readability, can draw an author's attention away from other important text characteristics. Therefore, we wanted to provide *diverse* measures so users would think about a constellation of features. For instance, one program delivers over forty measures of text characteristics, some measuring word characteristics, others measuring sentence characteristics.

We assume that authors have diverse needs, which cannot be foreseen entirely. Hence, the Writer's Workbench system contains many programs and options that can be used, or not, as an author desires. The diversity of programs works as a structured system to let users explore characteristics of words (such as syllables), sentences (such as length in words), or paragraphs (such as beginning and ending sentences).

2.3 Relative judgments

A text is good or bad depending on whether it is adequate for a particular task or audience. *Evaluative* judgments about a document usually relate to some standard of excellence. The Writer's Workbench system makes such comparisons explicit. Measures from one document are compared to measures obtained from other documents. For instance, evaluative comments about the readability of a text can be based on statistics derived from training or technical memoranda judged to be of high quality. The author is warned when the text being analyzed departs significantly from the standard documents. Ability to compare documents to various standards emulates an important component of human text judgments, and it is a distinctive feature of the Writer's Workbench system.

2.4 User modifications

Since it is not possible to anticipate the documentation needs of all users, programs should be *modifiable*. Three aspects of the programs can be adjusted by the user—standards, input, and output. For instance, one program calculates statistics for variables derived from texts supplied by an author. These statistics can then be used as standards to evaluate features of the author's subsequent texts, such as percentage of passive sentences. Other programs allow adding or deleting words or phrases that are detected by spelling and word usage dictionaries. In addition, input can be modified to include or exclude lists from analyses, and output format and length can be changed by appending simple descriptors to commands.

2.5 Text location

For authors to review exactly where language may have gone awry in a text, *specific* text locales must be identified. For instance, printing an average measure of sentence length is less useful for an author than printing sentences that exceed a specified length. Early work on identifying troublesome text locales for a reader was done by Koether and Coke.[14] Command options in the Writer's Workbench system provide summary statistics or review sentences that violate certain standards, such as length or word use. In the latter case, line numbers show where offending text can be found. This level of specificity is necessary if programs are to speed text revision.

2.6 Information resources

The Writer's Workbench system is intended to be *informative* rather than to make rigid decisions. Text statistics should be interpreted cautiously, and we suspect they will be used most effectively when an author can get advice related to those statistics. The Writer's Workbench system provides guidance, using information files, about word use, punctuation, and other features. Brief explanatory comments are also given in programs, with information about how to obtain more information.

III. RATIONALE AND DESIGN

"*UNIX*™ Writer's Workbench Software: Rationale and Design,"[11] describes program architecture, and discusses at length the literature behind programs and their options. Current literature on writing is rich in descriptions of text patterns that characterize faulty expression,[15] and contains descriptions of actions that a writer can take to correct those faults. These patterns and actions, which are the bases for program measures and recommended revisions, are described in the first part of the next paper. Next, the individual programs and their options are described. The programs are divided into those that proofread, those that comment on style, and those that comment on organization. Finally, features that help people use the programs—human factors issues—are surveyed.

IV. RESULTS OF A FIELD STUDY

The Writer's Workbench system has been used in Bell Laboratories for over three years; currently it runs on over 100 *UNIX* systems. Bell Laboratories users have responded favorably to the programs, and many believe the programs improve writing skills.[16]

Of course, there are many environments in which language analysis programs might be used. "*UNIX*™ Writer's Workbench Software: Results of a Field Study,"[17] describes data collected from program trials conducted outside Bell Laboratories. The two outside locations included one human factors organization and one technical writing organization. The groups were engaged in different writing tasks and the staffs had different computer experience.

According to the field study, authors liked the detailed computer suggestions about how to revise their texts, and they were better able to detect errors with program help than without it.

V. CONCLUSIONS

The Writer's Workbench programs do not do everything humans do when evaluating text. They cannot determine whether important content is missing, nor do they assess format features or the meaning of texts, as humans would. Nevertheless, the programs speed editing, and they help detect features that authors might otherwise miss.

Language analysis programs have many uses. Components of the Writer's Workbench system have been used for research; for instance, to derive general quantitative descriptions of text style and to relate these styles to text difficulty.[18] They have also been used, by various groups at Bell Laboratories, to help develop technical documents for new software systems and for writing technical memoranda. Programs could be used to obtain management information, for example, to monitor changes in documents across time, and to determine the match between skills of readers and the texts they must read.[19] Finally, the programs can be used for instruction to provide detailed feedback to students about the characteristics of their writing, or by an instructor to evaluate the effects of instruction on student writing.

The Writer's Workbench programs have recently been used as an adjunct to a technical writing course sponsored by the Bell Laboratories Education Center. Such programs could be especially useful to authors whose native language is other than English. Use of language analysis programs in research, development, management, and instruction will no doubt grow as these programs increase in diversity and sophistication.

VI. ACKNOWLEDGMENTS

Programs created by L. L. Cherry contributed much to the development of the Writer's Workbench system. She also helped with program enhancements. W. Vesterman, of Rutgers University, stimulated early work on stylistic and diction programs. S. A. Keenan made important contributions to program development and documentation, as well as to Writer's Workbench system trials. J. L. Collymore, M. L. Fox, and M. F. Poller also contributed to development, and exchanges with W. F. Fox and J. J. Dever were helpful.

REFERENCES

1. W. G. Muller, "Useability Research in the Navy," in T. G. Sticht and D. Zapf (Eds.), *Reading and Readability Research in the Armed Services,* Alexandria, VA.: Human Resources Research Organization, 1976.
2. "Plain Language Laws," in "Simply Stated," Newsletter of the Document Design Center, Washington, D.C.: American Institutes for Research, No. 18 (July, 1981), p. 1.
3. M. J. Rochkind, "The Source Code Control System," IEEE Trans. Software Eng., *se-1,* No. 4 (December 1975), pp. 364–70.
4. M. H. Bianchi, R. J. Glushko, and J. R. Mashey, "A Software/Documentation Development Environment Built from the UNIX Toolkit," Int. Fed. Inform. Processing, Working Conf. Automated Tools Inform. Syst. Design and Development (January 1982), New Orleans, LA.
5. J. Hartley, "Eighty Ways of Improving Instructional Text," IEEE Trans. Professional Commun., *PC-24,* No. 1 (March 1981), pp. 17–27.
6. E. S. Ferguson, "The Mind's Eye: Nonverbal Thought in Technology," Science, *197,* No. 4306 (August 1977), pp. 827–36.
7. R. A. Lanham, *Revising Prose,* New York: Charles Scribner's Sons, 1979.
8. L. E. McMahon, L. L. Cherry and R. Morris, "UNIX Time-Sharing System: Statistical Text Processing," B.S.T.J., *57,* No. 6 (July–August 1978), pp. 2137–54.
9. S. A. Keenan, "Computer Projections of the Cognitive Effects of Text Changes," Amer. Educ. Res. Assn., Boston, MA, April 7–11, 1980.
10. R. O. Duda and J. G. Gachnig, "Knowledge-Based Expert Systems Come of Age," Byte, 6, No. 9 (September 1981), pp. 238–81.
11. N. H. Macdonald, "*UNIX*™ Writer's Workbench Software: Rationale and Design," B.S.T.J., *62,* No. 6 (July–August 1983), pp. 1891–1908. Reprinted in this volume.
12. W. Strunk, Jr. and E. B. White, *The Elements of Style,* New York: Macmillan, 1972.
13. L. T. Frase, "Ethics of Imperfect Measures," IEEE Trans. Professional Commun., *PC-24,* No. 1 (March 1981), pp. 48–50.
14. M. E. Koether and E. U. Coke, "A Scheme for Text Analysis Using FORTRAN," Amer. Educ. Res. Assn., New Orleans, LA, February 25–March 1, 1973.
15. L. Flower, *Problem-Solving Strategies for Writing,* New York: Harcourt Brace Jovanovich, Inc., 1981.
16. L. T. Frase, N. H. Macdonald, P. S. Gingrich, S. A. Keenan and J. L. Collymore, "Computer Aids for Text Assessment and Writing Instruction," N.S.P.I. Journal, (November 1981), pp. 21–4.
17. P. S. Gingrich, "*UNIX*™ Writer's Workbench Software: Results of a Field Study," B.S.T.J., *62,* No. 6 (July–August 1983), pp. 1909–19.
18. E. U. Coke, unpublished work.
19. E. U. Coke and M. E. Koether, "A Study of the Match Between Technical Document Difficulty and the Reading Skills of Technical Personnel," B.S.T.J., *62,* No. 6 (July–August 1983), pp. 1849–64.

The *UNIX*™ Writer's Workbench Software: Rationale and Design

Nina H. MacDonald
AT&T Bell Laboratories, Murray Hill, New Jersey

The UNIX™ *Writer's Workbench software is a set of computer programs that help with two stages of document production: evaluation and editing. These programs analyze prose documents and suggest improvements. There are several types of programs: those that proofread, analyze style, and reformat the text in new ways, and those that provide information about the English language. This paper first describes the rhetorical and psychological writing principles that underlie the Writer's Workbench programs. It then describes the major Writer's Workbench programs and how they judge writing, based on these writing principles. Finally, it presents the human factors principles used in the design and development of the Writer's Workbench system.*

I. INTRODUCTION

The previous paper in this issue of the *Journal*[1] pointed out the growing need for automated language processing, that is, for tools to help authors write clearly and understandably. Such tools are especially important for technical writing where the content is precise and a reader's failure to understand the text can be costly.[2] This paper describes how the *UNIX** Writer's Workbench software attempts to meet this need.

The first major section of this paper describes the principles of good writing that are incorporated into the Writer's Workbench programs. Many of these principles are supported by research showing that, indeed, different writing styles make a difference to the reader. The next section describes the major Writer's Workbench programs and relates them to the principles. The last major section describes the user considerations that guided the design of the Writer's Workbench programs.

II. SOME PRINCIPLES OF GOOD WRITING STYLE

This section presents some principles of good writing style. Research suggests that text that violates these principles is more difficult to comprehend. These principles belong to the word, sentence, paragraph, or document level, but of course, some principles overlap categories.

The Bell System Technical Journal, Vol. 62, No. 6, July–August 1983, 1891–1908. Reprinted with permission from the AT&T Technical Journal.
*Trademark of Bell Laboratories.

2.1 Word level

There are several word level principles besides the obvious ones such as using and spelling words correctly.

2.1.1 Wordiness

Style guides usually advise writers to avoid hackneyed, empty, or frequently misused phrases, such as "at this point in time" and "notwithstanding the fact that."

2.1.2 Definite, specific, concrete language

Strunk and White[3] remark that "the surest way to arouse and hold the attention of the reader is by being specific, definite, and concrete." They illustrate this principle by this pair of sentences, the first vague, the second concrete:

A period of unfavorable weather set in.
It rained every day for a week.

Psychological research on memory and the imageability (ability to create an image) of words suggests that texts with many abstract words will be more difficult to remember, and presumably to understand, than texts with concrete words.[4–8]

2.1.3 Word frequency and length

Coleman reported a high negative correlation between the average frequency of content words in a passage and its difficulty to readers.[9] That is, texts made up of many infrequent words are difficult to comprehend. He also found that difficult passages contained more letters, syllables, and morphemes per word. Since frequency and length are highly correlated,[10] these two effects are possibly one effect. In keeping with these findings, Klare recommends Anglo-Saxon-based words over Latin-based words in English since the Latin form is usually longer, as in "go" versus "proceed."[11]

2.2 Sentence level

There are several obvious sentence-level principles, for instance, using correct punctuation and correct grammar. Some of the less obvious or more subjective principles are described below. These are not principles of right or wrong, but rather of better or worse.

2.2.1 Passive voice

One of the biggest writing problems, particularly in scientific writing, is the overuse of the passive voice. Historically the passive voice was used to indicate the objectivity of science and the scientist. The scientist did not state, "I found that . . . ," but rather, "It was found that. . . ." This usage was dogma; as Einstein said, "When a man is talking about scientific subjects, the little word 'I' should play no part in his exposition."[12]

One problem with the passive voice in scientific materials is that its use has spread from obscuring "I" and "we" to many other cases as well, e.g.,

A variable-gain control is included in this circuit.

Scientific objectivity is still served by stating:

This circuit includes a variable-gain control.

Perhaps because of a change in scientists' perceptions of their role,[13] or perhaps because of the difficulty of the passive style, many scientists and scientific editors[14,15] now recognize and even promote the use of the active voice, including the use of first-person pronouns.

Why should we avoid the passive voice? The rhetoric books describe it as "dron[ing] like nothing under the sun," wordy and unclear,[16] and less direct and less vigorous[3] than the active voice. It may indeed by all those things, but in addition, psychological research has shown that the active versions of a sentence are recalled better[17] and verified faster.[18] Scientific texts written in the third person passive, as "It was concluded that . . ." are remembered less well and appreciated less than the same content written in the active voice.[19]

Kirkman[20] took samples of scientific papers and rewrote the content in six different styles. He asked scientists and engineers to rate which version they found "most comfortable to read, easiest to grasp and simplest to digest." In three different surveys, he found they preferred "direct, active writing, with a minimum of specialist vocabulary, a judicious mixture of personal and impersonal constructions, short and uncomplicated sentences and liberal paragraphing."

These data do not imply that passive voice should never be used; at times it is preferable to

1. Emphasize the object of the sentence
2. Vary the rhythm of the text
3. Avoid naming an unimportant actor.

EXAMPLE: The mail was delivered.

However, the passive voice should be restricted to the useful and necessary cases, rather than used widely and indiscriminately.

2.2.2 Nominalizations

Nominalizations are nouns that have been created from verbs. They usually end in "ion," "ment," "ence," or "ance," e.g., "transformation," "establishment," and "admittance." The empirical case against using nominalizations is strong. Coleman[21] found that individual sentences without nominalizations were remembered better than their nominalized forms. A multiple-choice comprehension test on content failed to show a significant difference between the two versions. But on a memory task, subjects required significantly more exposure to memorize sentences containing two nominalizations than to memorize the same content written in active-verb form.

In a later experiment, Coleman[17] investigated ten different types of nominalizations and their active versions. He found that for those pairs of sentences in which the active version was phrased in two clauses and the nominalized version in one, the active was memorized more quickly. For example

ACTIVE: If he discusses the reason for the price-change, it will be appreciated.

NOMINALIZED: His discussion of the reason for the price-change will be appreciated.

Coleman[21] took passages from a psychology text and rewrote all the nominalizations, passive sentences, and adjectivalizations (verbs formed from adjectives) into active sentences. (Coleman[17] found no difference between adjectivalizations and their active forms.) He found that students answered correctly 25 percent more questions from the rewritten texts than from the originals. In a similar experiment, subjects were asked to write the passages immediately after reading them. Subjects recalled the simplified passages significantly better than the originals.

2.2.3 Expletives

In grammar, the term expletive refers to a syllable, word, or phrase that adds no information. In particular "expletives" are words such as "it" or "there," which anticipate a later word or phrase. Thus, in "There are three solutions to this puzzle," "There" is an expletive anticipating "solutions."

Many times such expletives can be deleted, e.g., "This puzzle has three solutions," although sometimes they cannot, e.g., "It is raining." Although no research demonstrates that expletives make text more difficult, Brogan[22] argues that when the expletive deemphasizes the main verb inappropriately, the sentence should be changed. For instance,

It is this necessity that adds to their complexity.

can be changed to

This necessity adds to their complexity.

making "adds" more salient.

2.3 Paragraph level

2.3.1 Readability

The readability or reading grade score for a text predicts how many years of schooling a reader would need to understand it. (Units other than years of schooling are sometimes used.) The prediction is usually based on the length of the words in the text and the length of the sentences. Different readability formulas calculate the lengths differently and weigh the factors differently.

As we mentioned in the previous section, the length of a word (highly related to its frequency) predicts its difficulty. Sentence length is related to sentence type, with complicated sentences usually containing more words than simple ones. Readability formulas predict reasonably well the difficulty of the text, not because sentence length and word length cause reading difficulty, but because they are highly correlated with features such as complexity and frequency, which do.

As with any predictor, these formulas can be fooled. The Dale-Chall formula[23] takes vocabulary items into consideration, but most formulas do not and will provide incorrect readability scores for nonsense text. All the formulas will give the same values for text with the sentences input backwards or forwards. But, in general, the formulas give a reasonable prediction of text difficulty when presented with naturally written text.

Unfortunately, research has shown[11] that the comprehensibility of the text is not necessarily improved, although the reading grade score is, by simply shortening words and sentences. The best way to improve the comprehensibility of a text is to rewrite it following the principles of good writing.

2.3.2 Variation

In writing, as in most fields of endeavor, moderation is best. The previously described principles are not absolutes; some passives and nominalizations are reasonable, and in fact, variation in sentence length, structure, and type is necessary to make writing interesting and keep the reader's attention.[13] There are other more important reasons to vary sentence type, which usually varies length as well.

Writing instructors suggest that less important ideas should be grammatically subordinated to more important ones so that the grammatical structure emphasizes the logical structure. Two simple sentences can be joined by using a "that" clause or an adverb, such as "although," to subordinate one to the other. The less important sentence should be in the subordinate clause after the "that" clause or adverb. For example, the following sentences:

1. The short, simple sentence is the most comprehensible form for an individual sentence.
2. Overusing such sentences may make a document seem disjointed.

Some sentences can be combined:

> Although the short, simple sentence is the most comprehensible form for an individual sentence, overusing such sentences may make a document seem disjointed.

The combined sentence subordinates sentence (1) to sentence (2), thus emphasizing that the information in sentence (2) is more important than that in sentence (1).

2.4 Document level

2.4.1 Organization

Most books on writing recommend that the first sentence of each paragraph present the topic of the paragraph or else provide a transition from the previous paragraph into a new topic.[3,13] If, for most paragraphs, the first sentence reflects the topic, then these sentences give the reader a reliable signpost to the meaning. Headings also provide signposts to topics and topic changes. A paper with good headings and topic sentences can be skimmed easily and quickly, and even the person who reads every word will find it easier to follow.

2.4.2 Audience considerations

Style books strongly advise writers to know their audience and to write for them. This is particularly important for materials such as instruction manuals, which the reader needs to understand. Writing for the reader extends from questions of vocabulary and sentence structure to content and organization. For content and organization, Flower[24] gives particularly thorough advice.

III. PROGRAMS

Until recently, students of writing could use only books and teachers to help them. This has slowly been changing with the advent of computer programs to do some of the work. Most programs, however, have focused solely on checking and correcting spelling.[25] Several readability indices have also been automated, but in general, wire services, magazines, newspapers, and businesses still do not analyze their text with the computer, although it is often stored and edited in a computer.

Although not yet in widespread use, several text analysis systems do exist. Besides the Writer's Workbench programs[26,27] to be described here, other systems include EPISTLE,[28] an IBM project still in the research stage; JOURNALISM,[29] a University of Michigan system that provides feedback to journalism students; and CRES,[30] a Navy system to help improve the quality of technical manuals and training materials. The EPISTLE system is a business office system that will abstract the contents of incoming letters but will also correct grammatical errors in outgoing letters. JOURNALISM provides some proofreading, but because it is programmed with specific knowledge about the articles it evaluates, it is able to comment on the organization and content as well. CRES calculates the readability score for the text, flags uncommon and misspelled words and long sentences, and suggests simple replacements for difficult words and phrases.

The rest of this section describes some of the Writer's Workbench programs, focusing on those that most strongly reflect the principles discussed in Section II.

3.1 Proofreading: *proofr*

The most useful Writer's Workbench programs are in some ways the least interesting. Every writer can use proofreading help, since as we become more familiar with a piece of writing we become poorer at spotting errors in it.

The proofreading program, *proofr,* invokes five separate programs. A three-line example of input text and its *proofr* output are shown in figure 1. Each of these five programs can be run individually, but are more conveniently run

```
INPUT:    Our report, "The Basic Fundamentals of Organizatonal Complexity",
          is enclosed.  Please send any recomended changes at your
          earliest convenience. thanks.


PROOFR    ***************************** SPELLING *****************************
OUTPUT:   Possible spelling errors in examplefile are:

          Organizatonal                 recomended

          If any of these words are spelled correctly, later type
                      spelladd word1 word2 ... wordn
          to have them added to your spelldict file.

          ************************** PUNCTUATION **************************

          The punctuation in examplefile is first described.

          2 double quotes and 0 single quotes
          0 apostrophes
          0 left parentheses and 0 right ones

          The program next prints any sentence that it thinks is incorrectly
          punctuated and follows it by its correction.

          line 1
          OLD: Our report, "The Basic Fundamentals of Organizatonal Complexity",
          NEW: Our report, "The Basic Fundamentals of Organizatonal Complexity,"
          line 3
          OLD: earliest convenience. thanks.
          NEW: earliest convenience. Thanks.

          For more information about punctuation rules, type:

                          punctrules

          ************************** DOUBLE WORDS **************************

          For file examplefile:

          No double words found

          ************************** WORD CHOICE **************************

          Sentences with possibly wordy or misused phrases are listed next,
          followed by suggested revisions.

          beginning line 1 examplefile
          Our report, "The *[ Basic Fundamentals]* of Organizatonal Complexity",
          is enclosed.

          beginning line 2 examplefile
          Please send any recomended changes *[ at your earliest convenience]*.

          file examplefile: number of lines 3, number of phrases found 2

          ---------------------------- Table of Substitutions ----------------------------

          PHRASE                        SUBSTITUTION

          at your earliest convenience: use "soon" for " at your earliest convenience"
          basic fundamentals: use "fundamentals" for " basic fundamentals"

          ************************** SPLIT INFINITIVES **************************

          For file examplefile:

          No split infinitives found
```

Figure 1. Input to and output from *proofr* program.

as a package. When run separately, some of the programs have capabilities not found when they are run as part of *proofr*. Each separate program will be discussed in turn.

3.1.1 Spelling: spellwwb

The *spellwwb* program, a spelling checker based on the *UNIX* system *spell* program,[31] allows users to have a personal dictionary of words. The program searches the input text and prints all words that are not in its dictionary or the user's dictionary.

In addition, the *spellwwb* program can be used interactively to correct spelling errors. The *spellwwb* program prompts the user with each possibly misspelled word. Responding to each, the user can:

1. Find all lines on which it appears in the file
2. Invoke another program to try to determine the correct spelling
3. Tell the program how to change the spelling
4. Leave it as it is
5. Save it in a personal dictionary of correct words.

The user can also specify that certain misspellings always be changed. For instance, a poor speller might store the correction "relevant" for "relevent." Then *spellwwb* makes all specified corrections and updates the user's personal spelling file.

3.1.2 Punctuation: punct

The *punct* program searches for simple punctuation errors. It recommends changes to:

1. Move commas and periods to the left of double quotes, and move semicolons and colons to the right of double quotes
2. Capitalize the first letter of sentences
3. Balance double and single quotes and parentheses.

The program enforces straightforward rules, not ones that require judgment, such as deciding whether a comma or semicolon is the appropriate mark. When *punct* finds an error, it prints the original line, followed by its correction of the line.

Some of the rules *punct* enforces are unfamiliar to many people, such as the relative position of double quotation marks with other marks of punctuation. Nevertheless, they are accepted standards of American English. The *punct* program directs a user who has made punctuation errors to use the program *punctrules,* which provides a list of pertinent punctuation rules. The user, thus, has easy access to reference information, which can be used to decide whether the suggested changes are appropriate.

3.1.3 Consecutive occurrences of the same word: double

Using context line editors for editing text increases the chance of having the same word twice in a row. The *double* program locates consecutive occurrences of the same word, which can be split across two lines.

3.1.4 Wordy phrasing: diction

The *diction* program, described by Cherry,[32] searches a text file for phrases that writing experts have classified as wordy or frequently misused. The latest version of its dictionary also contains some phrases that may reflect a sexual bias. The program prints sentences containing such phrases and surrounds them with stars and brackets (*[]*). It then recommends substitutions for these phrases. For instance, for the phrase, "bring to a conclusion," it recommends using "conclude," "end," or "finish."

Users should have the flexibility of tailoring the word list to their environment. In general, "end," "stop," or "finish" may be good choices for "terminate," but for many people writing about software, "terminate" is the appropriate word. Such a user can direct the program to stop searching for "terminate" by creating a personal dictionary. Having a private dictionary of extra terms and suppressed terms increases the usefulness of the *diction* program.

3.1.5 Split infinitives: splitinf

The *splitinf* program uses a "parts of speech" analysis program, *parts*,[33] to find infinitives that are split by adverbs. Such split infinitives as in "to quickly decide" are the most common type. Since users may be unfamiliar with this error, those whose papers contain split infinitives are told to use *splitrules,* which will print grammatical information about split infinitives.

3.2 Stylistic analyses

The stylistic analysis programs provide information whose interpretation is less concrete than that given by the proofreading programs. Hence, the information is more difficult to use; following the advice faithfully can require a considerable amount of time.

3.2.1 Tabular stylistic information: style

The *style* program,[32] based on *parts,* provides seventy-one numbers describing the stylistic features of a text. The most important variables it reports are several readability indices (described more fully by Cherry[32]), information on the average length of the words and sentences, the distribution of sentence lengths, the grammatical types of sentences used, e.g., simple and complex, the percentage of verbs in the passive voice, the percentage of nouns that are nominalizations, and the number of sentences that begin with expletives.

The *style* output for this article, shown in figure 2, is more useful for research on the style of documents, however, than for helping inexperienced writers improve their writing style. The *style* table is difficult for many writers to interpret because

1. Users may not know the meaning of some terms, e.g., "expletive" and "nominalization."
2. Users frequently do not know whether the numerical values should be high or low, even for terms that are probably familiar, such as "complex sentence."

```
readability grades:
        (Kincaid) 11.3 (auto) 12.6 (Coleman-Liau) 13.1 (Flesch) 13.2 (48.8)
sentence info:
        no. sent 240 no. wds 4636
        av sent leng 19.3 av word leng 5.18
        no. questions 1 no. imperatives 0
        no. content wds 2734 59.0% av leng 6.72
        short sent (<14) 24% (58) long sent (>29) 9% (22)
        longest sent 64 wds at sent 150; shortest sent 4 wds at sent 70
sentence types:
        simple 42% (101) complex 38% (92)
        compound 7% (16) compound-complex 13% (31)
word usage:
        verb types as % of total verbs
        tobe 32% (170) aux 16% (85) inf 17% (89)
        passives as % of non-inf verbs 14% (63)
        types as % of total
        prep 10.5% (487) conj 3.8% (177) adv 4.2% (197)
        noun 28.0% (1296) adj 17.2% (797) pron 4.7% (220)
        nominalizations 2 % (90)
sentence beginnings:
        subject opener: noun (48) pron (28) pos (1) adj (35) art (57) tot 70%
        prep 13% (32) adv 6% (15)
        verb 1% (3) sub_conj 6% (14) conj 2% (5)
        expletives 1% (2)
```

Figure 2. Style program output for this paper.

3. Users who know enough to minimize or maximize the use of some construction still do not know what numerical value is appropriate.

3.2.2 Interpreted stylistic analysis: prose

The *prose* program goes beyond the *style* program by providing the *style* statistics and an interpretation as well. The *prose* program compares the *style* values of the user's text against a set of standards and describes the differences in a two-to-three page output written in English sentences. A section of *prose* output is shown in figure 3.

Several sets of standards for comparison are available since texts are written for different types of readers and for different purposes. Users select which set of standards should be used in the interpretation of their text. Currently there are built-in standards for technical papers and prose training materials. These standards were set as follows. Department heads in the Bell Laboratories basic research area were asked to identify the best technical writers in their departments. These people were, in turn, asked to identify their best written technical documents (content aside). This process yielded twenty-eight documents. An editor in a Bell Laboratories training department provided thirty-four documents that he judged to be particularly well

SENTENCE STRUCTURE

Passives

This text contains a much higher percentage of passive verbs (44.0%) than is common in good documents of this type. The score for passive verbs should be below 28.6%. A sentence is in the passive voice when its grammatical subject is the receiver of the action.

PASSIVE: The ball was hit by the boy.

When the doer of the action in a sentence is the subject, the sentence is in the active voice.

ACTIVE: The boy hit the ball.

The passive voice is sometimes needed

1. to emphasize the object of the sentence,
2. to vary the rhythm of the text, or
3. to avoid naming an unimportant actor.

EXAMPLE: The appropriations were approved.

Although passive sentences are sometimes needed, psychological research has shown that they are harder to comprehend than active sentences. Because of this, you should transform as many of your passives to actives as possible. You can use the *style* program to find all your sentences with passive verbs in them, by typing the following command when this program is finished.

style -p filename

Nominalizations

You have appropriately limited your nominalizations (nouns made from verbs, e.g., "description").

Figure 3. Section of *prose* program output for a poor paper.

written. The technical and training documents were then run through the *style* program. The means and standard deviations of each of the *style* variables were computed for each document set. These means and standard deviations now make up the standard for that document type. When the value of a user's variable is more than one standard deviation from the mean of that standard, *prose* recommends changes. The *mkstand* program can be used to create additional standards from a set of documents. This flexibility allows any writing group to tailor the standards for its particular audience.

If any of the input text's values is more than one standard deviation from the selected standard, *prose* explains why this may make the document hard to comprehend and how to rewrite the text to remove the problem. If scores on a variable are acceptable, and the variable is an important one, *prose* tells the user that the text achieved an appropriate score.

3.2.3 Stylistic problems in context: findbe

Although the *style* program can isolate individual sentences that contain passive verbs, expletives, or nominalizations, it is frequently difficult to know which sentences to change and how to change them without seeing the surrounding context. The *findbe* program automates part of a prescription for revision given by Lanham, who tells the revisor to circle all forms of the verb "to be" and to try to replace them.[34,35] This advice agrees with some of the information presented earlier since "to be" occurs with passives, e.g.,

The difficulty of the passive voice IS *disregarded* by many writers.

with many nominalizations, e.g.,

Coleman's *discussion* of nominalizations *IS* comprehensive.

and with expletives, e.g.,

There ARE many ways to avoid expletives.

The *findbe* program underlines and capitalizes all forms of "to be." The text is then formatted as usual, providing the user with a paper that is easy to read, since all sentences are in context and problem areas are highlighted. This turns out to be a useful way of looking at the first draft of a paper.

3.2.4 Checking text abstractness: abst

The *abst* program[36] indexes the conceptual abstractness of the input text. Abstractness is defined here as the percentage of words in the text that also occur on a list of 314 words rated as abstract in psychological research.

When the percentage of abstract words is over 2.3%, the program suggests that concrete examples be introduced to make the document more understandable. (This cutoff was derived from the collection of good documents used to develop the *prose* standards.) The abstract words found in the text are saved in a file for the user to review.

3.3 Organization

3.3.1 Judging organization: org

The organization of a text is important since an appropriate structure makes it more comprehensible. For a computer program to analyze the organization of a text fully, it would need to abstract the content. Without a parser for English, or some other way of interpreting the meaning of a text, our programs cannot give feedback on the quality of the content and organization.

The *org* program, however, was designed to give a writer a different perspective on a text as an aid in evaluating its organization. The program formats the text and preserves headings and paragraph boundaries, but prints only the first and last sentence of each paragraph.

For writers who follow the traditional topic sentence and concluding sentence format for paragraphs, the output can be a good abstract of the paper. Even for writers with a more casual style, seeing the overall structure of a long paper can help to improve it.

IV. USER CONSIDERATIONS

In discussing human-computer interactions, two issues should be considered: the ease of using the programs and the quality of the computer's responses.[37] This section will describe how the Writer's Workbench system attempts to optimize both aspects of the interaction.

4.1 User interactions

4.1.1 Program power

Perhaps the most important organizational decision made was to design the Writer's Workbench system hierarchically. One command, *wwb*, runs the most commonly used programs, the proofreading program, *proofr*, and the English-language stylistic program, *prose*. The *wwb* program is easily remembered as the acronym for the Writer's Workbench system, and so the casual user of the system finds the system simple to access.

For the experienced user of the Writer's Workbench programs, *proofr* and *prose* can be used individually, as can the separate components of *proofr*, thus allowing such users all the power they need. For one program, *proofr*, there is an alternate spelling, *proofer*, because based on the pronunciation it was often misspelled this way.

Many of the programs allow users to have their own dictionaries to tailor the output. Rather than requiring the users to type the names of the dictionaries as part of the command line, these programs use a particular file if it exists. Of course, users can still override this when they choose. Furthermore, commands were created to add words to these dictionaries, rather than requiring the users to edit them and keep them in sorted order. Thus, casual users can create personal spelling dictionaries and access them automatically with just two commands.

4.1.2 Documentation

For new users, there is now a substantial amount of paper documentation that describes the programs, how to use them, and their relation to good writing. But because users frequently do not own the paper documentation or do not have it near them, there are many on-line aids as well.

As with all *UNIX* system commands there are manual pages for each Writer's Workbench program, which are helpful to experienced *UNIX* system users. For the casual user, there is an on-line introductory system, which describes the uses of each Writer's Workbench program command.

Two other commands give on-line help. The *wwbinfo* command provides a list of all the commands and their functions. The *wwbhelp* command takes a key word as an argument and lists all the programs that have anything to do with that key word. These are words such as the following: "syllable," "prose," "passive," "sentence," and "organization."

Each Writer's Workbench program can also be run with a "flags" option, which prints that program's format and options. Further, since many of the programs have default options, these options are also echoed back with the user-specified options, as described in Section 4.2.2.

Finally, the output of many of the programs suggests other programs that would be useful. These suggestions are based on the analyses of the input text, and can thus remind users of programs that are useful for that text. For instance, users with spelling errors are told how to add correctly spelled words to their personal spelling dictionary, and the *prose* output suggests other appropriate programs. For example, users with too many passives are told how to find all sentences with passive verbs by using the *style* program.

4.2 Computer responses

What users have to remember and what they have to type are certainly important variables to consider when evaluating a system. But the quality of the messages the computer provides is also important.

4.2.1 Output length

The Writer's Workbench programs have attracted many new *UNIX* system users. For such users, the traditional "silence" of *UNIX* system commands is unfriendly,[38] e.g., if the *spell* command finds no misspelled words it simply stops, and the user receives a prompt (not a pat on the back). This silence is exactly what the regular *UNIX* system user wants, as McIlroy states, "Canned chitchat, like the plastic announcements on airplanes, may please newcomers, but it annoys old hands."[39] The obvious problem is that regular and casual users share machines, and what is right for one is frustrating for the other.

The first versions of most of the Writer's Workbench programs were verbose, irritatingly so, for experienced *UNIX* system users. We added "-s" options (for short) to most programs, which removed the chitchat and most of the "pats on the back." Recently, we changed all the programs so that users can specify once what length of output they want in the future. Since expert users are best able to change this default, users who do not specify any length receive the long version.

4.2.2 Feedback

The Writer's Workbench programs inform the user when the program has started. Even experienced users may want reassurance that everything is proceeding on a heavily loaded system. Most commands echo the command line and include any options that were not chosen by the user but came about through default. This gives users a record of the exact command run; it may also remind them that they do not want a particular default, and it may alert them to options they did not know existed.

The Writer's Workbench programs cannot correct incorrect entries, but they try to provide complete, informative and accurate error messages.

4.3 Needed improvements

To date, *proofr* is a major proofreading package, geared toward users who print the output on paper and make any recommended changes themselves. Since the output refers to a text problem by labeling it with the line number in the unformatted file, a user who has someone else type the text and only has the formatted output does not find these line numbers useful. For users with CRT display terminals, the output can be too lengthy to fit in the terminal's memory so that by the end of the program the first part of the output has disappeared. This makes it difficult to notice all the problems and change them.

We are currently designing two new versions of the *proofr* program. A version for word processing center customers will provide all the proofreading comments superimposed on the formattd output. For CRT users, we plan a completely interactive version, which will move linearly through the file and display possible corrections for the user to accept or ignore.

V. SUMMARY

This paper described some principles of good writing and some experimental results that show that readers prefer writing that embodies these principles and find it easier to understand. The paper then described a set of computer programs called the *UNIX* Writer's Workbench software. These programs help a writer isolate problems with general style as well as with individual sentences, phrases, and words. The final section of the paper described some of the human factors principles that guided the design of the programs.

REFERENCES

1. L. T. Frase, "The *UNIX*™ Writer's Workbench Software: Philosophy," B.S.T.J., *62,* No. 6 (July–August 1983), pp. 1891–1908. Reprinted in this volume.
2. J. P. Kincaid, R. P. Fishburne, R. L. Rogers, and B. S. Chissom, "Derivation of New Readability Formulas (Automated Readability Index, Fog Count, and Flesch Reading Ease Formula) for Navy Enlisted Personnel," Navy Training Command Research Branch Report 8-75 (February 1975).
3. W. Strunk, Jr., and E. B. White, *The Elements of Style,* New York: Macmillan Co., 1959.
4. A. Paivio, "A Factor-Analytic Study of Word Attributes and Verbal Learning," *J. Verbal Learning and Verbal Behavior, 7,* No. 1 (February 1968), pp. 41–49.
5. G. Frincke, "Word Characteristics, Associative-Relatedness, and the Free-Recall of Nouns," *J. Verb. Learn. Verb. Behav., 7,* No. 2 (April 1968), pp. 366–72.
6. W. A. Winnick and K. Kressel, "Tachistoscopic Recognition Thresholds, Paired-Associate Learning, and Immediate Recall as a Function of Abstractness-Concreteness and Word Frequency," *J. Exp. Psychol., 70,* No. 2 (August 1965), pp. 163–68.
7. A. M. Gorman, "Recognition Memory for Nouns as a Function of Abstractness and Frequency," *J. Exp. Psychol., 61,* No. 1 (January 1961), pp. 23–29.

8. R. C. Anderson and R. W. Kulhavy, "Imagery and Prose Learning," *J. Ed. Psychol., 63,* No. 3 (June 1972), pp. 242–43.
9. E. B. Coleman, "Developing a Technology of Written Instruction: Some Determiners of the Complexity of Prose," In E. Z. Rothkopf and P. E. Johnson (Eds.), *Verbal Learning Research and the Technology of Written Instruction,* New York: Teachers College Press, 1971, pp. 155–204.
10. G. K. Zipf, *Human Behavior and the Principle of Least Effort,* Cambridge, MA: Addison-Wesley, 1949.
11. G. R. Klare, *A Manual for Readable Writing,* Glen Burnie, MD: Rem Company, 1953.
12. A. Einstein, *Essays in Science,* New York: The Philosophical Library, 1949, cited in Mills and Walter, p. 29.
13. G. H. Mills and J. A. Walter, *Technical Writing,* 4th ed., New York: Holt, Rinehart, and Winston, 1962.
14. R. A. Day, *How to Write and Publish a Scientific Paper,* Philadelphia: ISI Press, 1979, p. 119.
15. M. O'Connor, *The Scientist as Editor: Guidelines for Editors of Books and Journals.* New York: John Wiley and Sons, Inc., 1979, p. 163.
16. S. Baker, *The Practical Stylist,* New York: Thomas Y. Crowell Co., 1962.
17. E. B. Coleman, "Learning of Prose Written in Four Grammatical Transformations," *J. Appl. Psych., 49,* No. 5 (October 1965), pp. 332–41.
18. P. B. Gough, "Grammatical Transformations and Speed of Understanding," *J. Verb. Learn. Verb. Behav., 4,* No. 2 (April 1965), pp. 107–11.
19. R. D. Ramsey, "Grammatical Voice and Person in Technical Writing: Results of a Survey," *J. Tech. Writing Comm., 10,* No. 2 (1980), pp. 109–13.
20. J. Kirkman, *Good Style for Scientific and Engineering Writing,* London: Pitman, 1980.
21. E. B. Coleman, "The Comprehensibility of Several Grammatical Transformations," *J. Appl. Psychol., 48,* No. 3 (June 1964), pp. 186–90.
22. J. A. Brogan, *Clear Technical Writing,* New York: McGraw-Hill, Inc., 1973, pp. 147–48.
23. E. Dale and J. S. Chall, "A Formula for Predicting Readability: Instructions," *Educ. Res. Bull., 27,* No. 1 (January 1948), pp. 37–54.
24. L. Flower, *Problem-Solving Strategies for Writing,* New York: Harcourt Brace Jovanovich, Inc., 1981.
25. J. L. Peterson, "Computer Programs for Detecting and Correcting Spelling Errors," *Commun. ACM, 23,* No. 12 (December 1980), pp. 676–87.
26. N. H. Macdonald, "Pattern Matching and Language Analyses as Editing Supports," *Amer. Educ. Res. Assn.,* Boston, April 1980.
27. N. H. Macdonald, L. T. Frase, P. S. Gingrich, and S. A. Keenan, "Writer's Workbench: Computer Aids for Text Analysis," *IEEE Trans. Commun.,* Special Issue on Communications in the Automated Office, *30,* No. 1 (January 1982), pp. 105–10.
28. G. E. Heidorn, K. Jensen, L. A. Miller, R. J. Byrd, and M. S. Chodorow, "The EPISTLE Text-Critiquing System," *IBM System J., 21,* No. 3 (1982), pp. 305–26.
29. R. L. Bishop, "The JOURNALISM Programs: Help for the Weary Writer," *Creative Computing, 1,* No. 2 (January/February 1975), pp. 28–30.
30. J. P. Kincaid, J. A. Aagard, J. W. O'Hara, and L. K. Cottrell, "Computer Readability Editing System," *IEEE Trans. on Professional Commun. PC-24,* No. 1 (March 1981), pp. 38–41.
31. M. D. McIlroy, "Development of a Spelling List," *IEEE Trans. Commun.,* Special Is-

sue on Communications in the Automated Office, *30,* No. 1 (January 1982), pp. 91–99.

32. L. L. Cherry, "Writing Tools," *IEEE Trans. Commun.,* Special Issue on Communications in the Automated Office, *30,* No. 1 (January 1982), pp. 100–105.
33. L. L. Cherry, "PARTS—A System for Assigning Word Classes to English Text," Computing Science Technical Report, *81,* Bell Laboratories, Murray Hill, N.J. 07974, 1978.
34. R. A. Lanham, *Revising Prose,* New York: Charles Scribner's Sons, 1979.
35. R. A. Lanham, *Revising Business Prose,* New York: Charles Scribner's Sons, 1981.
36. L. T. Frase, P. S. Gingrich, and S. A. Keenan, "Computer Content Analysis and Writing Instruction," Amer. Educ. Res. Assn., Los Angeles, CA, April 17, 1981.
37. R. W. Bailey, *Human Performance Engineering: A Guide for System Designers,* New York: Prentice Hall, 1982.
38. D. A. Norman, "The Trouble with UNIX," *Datamation, 27,* No. 12 (November 1981), pp. 139–50.
39. M. D. McIlroy, unpublished paper.

Plain Language through Protocol-Aided Revision

Karen A. Schriver
Carnegie Mellon University, Pittsburgh

Critics of the "plain language movement" point out that what is "plain" to one audience may mystify and confuse another. Questions such as "Plain language for whom?" and "How can we know whether a text is written in plain language?" raise legitimate concerns about the danger of ignoring that what is "plain" is relative to the particular audience reading and/or using a text. This paper addresses these questions directly by illustrating a concrete and empirically tested procedure for revising texts to meet the needs of particular audiences. Specifically, this paper details "protocol-aided revision," a procedure which employs reader responses to texts to guide revision activity, and demonstrates why actual reader feedback is the most sensible and effective criterion for deciding whether a text is written in plain language. It provides a case study of a text that was revised for comprehensibility using protocol-aided revision, underscoring that plain language is more than verbal text alone—it includes effective integration of text and graphics. In addition, this paper presents a model of the process of protocol-aided revision. This paper may interest both proponents and critics of the plain language movement, because it demonstrates that plain language is possible and that there are effective methods for clarifying text for particular audiences.

PROBLEMS IN DEFINING PLAIN LANGUAGE

In 1979, Veda Charrow asked proponents of the plain language movement, "What is plain English, anyway?" Charrow answers this question indirectly, asserting that "although most of us would probably not agree on what plain English is, we could probably agree on some aspects of what it is not" (Charrow 1979: 2). Charrow explicates the variety of ways that advocates have defined plain English, citing that it is not "legalese," or "plain, ordinary peoples' language," or "texts that are written with short sentences and simple words" (Charrow: 3). Early advocates such as Charrow spent a great deal of time trying to address the needs of at least four distinct audiences for plain language: (1) consumers, (2) critics, (3) government officials, and (4) writing and reading researchers. Each audience had a different set of concerns regarding a definition. Consumers wanted it to mean that they would be able to understand and use the documents they read and sign. Critics wondered if plain language meant "elimination of all technical words" and worried about its implications in the marketplace, its ramifications for creating and revising public documents, and its effect on texts that were intended for both novices and experts (MacNeil/Lehrer Report 1978: 4–5). Government officials wanted to know what a plain language law would look like and what enforcement would involve. Writing and reading researchers wanted to know if there were ways to verify whether a text was written in plain language; they wanted to move plain language beyond the slogan stage and to create a research agenda involving how people read and understand functional documents (Felker 1980).

At the same time that advocates were trying to define plain language, they were also trying to educate consumers. During the late 1970s, advocates of plain language were trying to inform the public that confusing and hard to understand public documents did not have to be the norm, that there were alternatives to medicalese, legalese, and bureaucratese. Consequently, most effort in the late 1970s in Britain and in the United States was devoted toward raising the consciousness of consumers that they indeed have a right to demand plain language. Early advocates provoked controversy by arguing that unclear and purposefully vague and jargon-laden language was being used as a tool to keep the less knowledgeable, less powerful, less wealthy, from knowing what they were signing.

When plain language came to the public's attention, there were very few publications concerning the nature of document design and writing processes (For a review of the recent literature on document design, see Schriver, 1989a.). There was also very little consolidated research on how people read and use public documents. Part of the mission of publications such as *The Information Design Journal, Visible Language,* and the American Institutes of Research's newsletter, *Fine Print* (now *Simply Stated*), and Carnegie Mellon University's *Communications Design Center Newsletter* was to inform researchers, educators, and practitioners of new findings on document design research as well as to provide a forum for raising plain language issues. However, before document design research had time to clarify the goals of plain language, thus allowing it to be defined more convincingly, lawyers and government officials were wrangling over its definition on American television (see, for example, The MacNeil/Lehrer Report, "Plain English" 1978).

It is unfortunate that plain English got off to such a confused start. Almost every country which has advocated plain language has found itself in the position of needing to clarify what it is and is not. The following characterization of plain language created by Australia's Law Reform Commission of Victoria (1986) demonstrates the continued need to clarify the nature of plain English:

> Plain English is a full version of the language, using the patterns of normal, adult English. It is not a type of basic English, or baby-talk. While documents that are converted to plain English may be described as *simplified,* they are simplified in the sense of being rid of entangled, convoluted language—language that is difficult to analyze and understand, language that submerges, confuses, and conceals its message. They are simplified in this sense, and not in the sense that the language has been severely condensed or amputated and the message truncated. Plain English is not artificially complicated, but it is clear and effective for

its intended audience. While it shuns the antiquated and inflated word, which can readily be either omitted altogether or replaced with a more useful substitute, it does not seek to rid documents of terms which express important distinctions. Nonetheless, plain language documents offer nonexpert readers some assistance in coping with these technical terms. To a far larger extent, plain language is concerned with matters of sentence and paragraph structure, with organization and design, where so many of the hindrances to clear expression originate (Law Reform Commission of Victoria 1986: 3).

This characterization makes two important points about the goals of plain language. First, to meet goals of plain language does not mean condescending to the reader. Second, writers of plain language documents intend neither to eliminate nor to hide complex ideas or technical terms in documents. These points are important ones that need to be reinforced and demonstrated before the goals of plain language will be understood by its critics.

Up to this point, however, plain language proponents have not sufficiently demonstrated methods for revising text that illustrate how plain language can be achieved. The purpose of this paper is to suggest that plain language can be defined and that specific methods exist for insuring a text's clarity for its intended readership. This paper will explain why collecting readers' responses to text is the most effective test for judging whether it is plain for the intended audience. Specifically, "protocol-aided revision," a procedure which uses reader feedback to guide revision of texts for comprehensibility and usability, is detailed. To show how protocol-aided revision works, a case study of the process is presented. In addition, this paper provides a model of the protocol-aided revision process, illustrating how protocols can help writers modify text to meet the special needs of particular audiences.

THE IMPORTANCE OF READER FEEDBACK

Since the 1940s, writers have been looking for ways to verify the success of public documents, and many techniques have been developed to aid the text evaluation process. The most widely used techniques, readability formulas such as those of Flesch (1949) and Gunning (1952), rely on counting the number of words and syllables per sentence to determine a text's readability. These techniques are by definition text-based; that is, they focus solely on surface-level text features and not on how readers respond to the texts. Tests of this sort have been shown to have severe limitations for guiding the revision of texts that are effective and usable (Duffy 1985). While such tests can provide gross clues suggesting which sentence-level features may be problematic, their output provides little, if any, information about how the text is working at the paragraph and whole-text level. In fact, when such text-based tests are used as the only guide for revision, revisers may actually make the text worse instead of better (Swaney, Janik, Bond and Hayes 1981).

Successful revision has been shown to depend on the writer's ability to anticipate the needs of a reader and to identify ways to help clarify whole-text problems from the reader's perspective (Sommers 1980; Schriver 1987; Hayes, Flower, Schriver, Stratman, and Carey 1987). To revise a text for a particular audience requires that writers recognize and predict the diverse goals and reading strategies that people may bring to the process of understanding and using a functional document.

Readers have been found to bring a wide variety of goals, purposes, assumptions, and reading strategies to their comprehension and use of functional texts. Some of the goals and purposes readers may bring include reading to:

- do a task, e.g., filling out a form for a loan application;
- understand an idea, a concept, or a definition, e.g., reading to understand one's rights in a legal contract;
- find information quickly, e.g., trying to find a procedure in a users' manual for operating a computer;
- assess the relevance of a text, e.g., reading a brochure which describes the conditions for a tax rebate;
- interpret and use the information for a purpose other than the text's intended purpose, e.g., reading a computer manual to solve a problem that is not described in the text, but that may be solved by analogous means;
- refresh one's memory about a fact, procedure, or idea that is vaguely remembered, e.g., reading a reference manual for a telephone answering machine to retrieve a fact about remote dialing;
- make a decision about choices or alternative ways to consider the same idea, e.g., reading a pamphlet that describes the pros and cons of building a nuclear power plant.

To determine if a text is meeting the goals and purposes of the intended audience, writers need more feedback than text-based tests can provide. Writers have found standard advice too simplistic to guide the revision of the complex texts they create. Writers want more than vague maxims such as "choose a suitable design and hold to it" and "avoid fancy words" (Strunk and White 1979: 15, 76).

Today's writers are often faced with a range of decisions, most of which are not at the sentence-level—decisions such as whether to present the text on paper or in some other media, for instance, video or online; decisions such as how to organize the text to promote rapid retrieval of information; or decisions such as how to choose optimal graphic devices to help clarify the text's structure and meaning. Instead of abstract "elements of style," writers are looking for answers to concrete questions concerning how well the text is functioning for the intended audience.

As audiences become more specialized and more educated in technical areas, they expect texts that are not designed "for everyone," but rather, are targeted to their particular needs. In many industrial and corporate document design contexts (such as the computer, electronics, and appliance industries), writers must tailor their texts to very particular audiences. The ability to adapt texts for audiences such as novices, intermediates, or experts is rapidly becoming a requisite skill in industry.

Writers are finding that the best way to evaluate the success of a functional document is perhaps the most obvious: observe readers while they try to understand and use the text. An effective revision procedure that uses the feedback of readers while they are engaged in the process of comprehending and using a text is "protocol-aided revision," a method developed in 1980 by researchers at Carnegie Mellon University's Communications Design Center.

WHAT IS PROTOCOL-AIDED REVISION?

Although protocol analysis as a method for studying cognitive processes has been widely discussed in the literature (for an excellent review, see Ericsson and Simon 1984), the use of protocols to guide reader-based revision of functional texts has not. Protocol-aided revision is a method for helping writers see problems in text that they might otherwise miss. It involves using readers' comprehension and performance difficulties as the basis for revision activity. It is a cyclical activity in which each cycle consists of readers responding to a text and a writer using readers' responses to guide revision. The next six sections of this paper are intended to help the reader understand how to use protocols in revision. These sections will cover the following topics: defining protocols and their functions, designing protocol tasks, selecting participants for protocol tasks, considering practical issues, collecting protocols, and finally, fixing problems uncovered by participants.

The nature of protocols

A protocol is a record of events, thoughts, or behaviors that occur over a period of time. Protocol analysis is a method for tracing a person's thinking or performance on a task. The record is usually obtained by using a videotape, an audiotape, or a computer program which records what a person says or does as he or she completes the task. Psychologists, decision-making theorists, writing researchers, and document designers are among those who have used protocol analysis to study how people think as they engage in activities such as solving chess problems (Simon and Gilmartin 1973), making decisions in supermarkets (Payne and Ragsdale 1978), planning, writing, or revising text (Hayes and Flower 1980; Berkenkotter 1983; Schriver 1987, 1989b; Hayes et al. 1987; Smagorinsky, 1989), or using computer manuals (Schriver 1984; Bond 1985; Sullivan and Flower, 1986; Schriver et al. 1986).

Generally, there are two categories of protocols which have relevance to writers: behavior protocols and thinking-aloud protocols. In behavior (or motor) protocols, the writer watches participants as they interact with and use texts such as forms, contracts, procedural instructions, or computer documentation, recording their actions and behaviors. The primary feature of this type of protocol is that participants do not talk aloud while performing a task—they simply do the task while either a writer or a computer program records what they do. Typically, writers collecting behavior protocols are interested in such issues as: where readers look for information (such as in indexes, in tables of contents, in glossaries); how quickly people can find information (as in searching for information online); how users make errors and recover from them while operating machinery; how features such as color, windowing, or display rate influence people's ability to use computers; or how quickly and accurately people can perform a task while using a set of instructions (for example, using a manual to assemble a bicycle).

Behavior protocols include eye movement studies, keystroke logs, and user-edits. Eye movement studies have been used by writers to determine the effect of colors, display rate, and cursor movement in online documentation and interface design. Keystroke logs, which can be collected automatically during interaction with a computer, provide detailed information about users' error and error-recovery patterns. User-edits, first described by Atlas (1981) involve having users try to perform a task (usually one which involves using a machine, equipment, consumer product, or device), using only its instruction manual as a guide.

In thinking-aloud protocols, on the other hand, participants are asked to perform a task while thinking aloud as they interact with a text and/or with a machine. When people experience difficulty in comprehending or in using the text, their comments typically reveal the location and nature of the difficulty. Unlike participants in behavior protocols, think-aloud participants are asked to verbalize anything that comes to their mind as they are engaged in the task. Because thinking-aloud protocols are collected while the person is reading and is engaged in the process of comprehension, they provide much more explicit and complete information than do readers' comments collected after reading is finished.

Typically, the constraints of practical situations require writers to choose between these two types of protocols, depending on their goals and purposes in evaluating a text. Since each type of protocol gives the writer a different "window" on the text, it is important to clarify one's purpose in evaluating the text before selecting a method. Behavior protocols are often employed when the purpose of evaluation is to determine how quickly and accurately people can use a text. Think-alouds are frequently used when the goal is to assess how people understand, solve problems with, draw inferences about, use, or read texts.

Writers interested in *where, how, and why readers make errors in using or in understanding a text* will find only the grossest clue for answering these questions with behavior protocols. A person using a computer, for instance, may make an error in issuing a command—a behavior that can easily be recorded with a keystroke protocol. The writer, however, needs to know "why" the error occurred. Did the person make the error because of: (1) a slip of the finger, (2) a misreading of the correct command, (3) a missing instruction, (4) a poorly worded instruction, (5) a faulty inference about what the instruction meant, (6) a misunderstanding of an example within the instruction, or (7) inaccurate, incomplete, or misleading feedback from the system itself? The advantage of think-alouds is that participants often say how and why they are having a difficulty with the text. Therefore, the writer has both locative and diagnostic information that will help guide revision decisions.

Over the past eight years, many document designers, teachers, and researchers have asserted that think-aloud protocols are one of the most sensitive ways to evaluate the quality of a functional document (Bond et al. 1980; Winbush and McDowell 1980; Atlas 1981; Swaney, Janik, Bond, and Hayes 1981; Roberts and Sullivan 1984; Bond 1985; Winkler, Ferguson, and Youngquist 1985; Lund 1985; Schriver 1984, 1987, 1989b; Sullivan and Flower 1986; Schriver et al. 1986; Witman 1987). For a discussion of theoretical and practical considerations in choosing text-evaluation methods, see Schriver (1989b).

Protocol-aided revision is now a widely used evaluation method in many human factors and testing labs across the country (Lewis 1983). Soderston (1985), for example, provides a detailed account of the usability edit conducted at IBM Kingston's Human Factors Laboratory; she argues that this procedure consistently reveals gaps and ambiguities in texts that have already gone through many technical reviews. Similarly, Lund (1985), from the Control Data Corporation, describes using "the candid camera approach" for evaluating the user interface of a new interactive graphics application. Moreover, protocol-aided revision, which employs think-alouds to isolate problem areas in text, is an extremely effective means for revising for comprehensibility.

Designing an evaluation using protocol-aided revision

Before conducting protocol-aided revision, writers will need to specify just what is expected from participants; that is, they will need to specify what participants will read or do. A primary consideration is to select a task (or tasks) which will best evaluate whether the text's goals for the audience are being met. For example, if the text is a reference manual, the task should test how successfully people are able to retrieve information from sections such as the index or the table of contents and then find the information in the main body of the text.

In addition to deciding the participants' task, writers must determine whether participants will read all of the text or just sections of it. If the text is too lengthy to conduct a complete evaluation, it is usual to isolate a section (or sections) to be evaluated. In so doing, it is important to select portions of the text that may be most sensitive to revealing how it is functioning for the audience. It is useful to evaluate the least complex sections of the text as well as the most difficult. Furthermore, if the text has a table of contents or an index, these sections should be tested. This will allow the writer to see how people respond to the varying degrees of difficulty within the text as well as to the text's access features. Normally, the task should take the participant no more than one hour to complete; otherwise the test results may be altered by the participant's fatigue.

In creating a context for the task, many writers use the "scenario approach," which involves giving participants a concrete goal for doing the task. For example, suppose the writer of a medical brochure describing the advantages and disadvantages of various surgical procedures wanted to know if readers clearly understood their options. The scenario provided to protocol participants might be "Imagine you are about to make an important decision about which surgical procedure to choose for your mother. Your goal in reading this brochure is to determine what decision to make based on the information provided." Usually, providing participants with a concrete goal or purpose will lead them to take an active role in reading and understanding the text. When participants have no goal or purpose in reading the text, they sometimes take a more passive role in their reading and understanding, tending to monitor their comprehension less often, thus verbalizing very little.

Selecting participants for the protocol task

To evaluate a text for its intended audience requires that participants in protocol-aided revision be members of (or at least share a great deal in common) with the intended audience. Before selecting participants for a protocol task, one should create a profile of the intended audience, for example, age, experience level, background, reading ability, attitudes, knowledge of technology, and so on. In general, it is a good idea to determine how much the participant knows about the topic as well as the participant's attitudes, preferences, and biases about the topic.

One of the most frequent questions writers ask about protocol-aided revision is "How many participants should I recruit?" While there are no definitive answers to this question, a few suggestions can be made. First, one should keep in mind that the goal in recruiting participants is to gather a variety of responses to the text rather than to ensure statistical reliability. It is important not to confuse protocol-aided revision with an experiment—the goal is not hypothesis testing. Rather, it is aimed at debugging poorly written text. Second, while collecting even one protocol is better than no protocol, that one may be highly idiosyncratic and unrepresentative of the larger intended audience. The important question is "how small can your participant sample be and still be useful?" In practice, document designers at the Communications Design Center have found that five participants per cycle has proved a useful number in conducting protocol-aided revision.

While a single cycle of protocol-aided revision is typically very helpful in comparison to other revision techniques, it is wise in most cases to use several cycles. A rough rule of thumb is that the first pass finds about half of the reader's problems, the second pass half of the remaining problems, and so on. Most texts can be revised to meet the reader's needs in two or three cycles. For each revision cycle, five new participants should be recruited. One should avoid asking the same person to provide a protocol on a text (or task) more than once.

Preparing to collect a protocol: some practical issues

To collect a protocol, you will need the following:

1. A set of clearly written instructions that can be given to each participant. Writers should note that the instructions can make or break your protocol testing. Be certain to pilot test your instructions for clarity with at least two people before conducting a formal protocol. The important question is: "Do participants interpret the task as I planned them to?" For examples of typical instructions for collecting protocols in document design studies, see Schriver (in press).
2. Recording equipment (e.g., audio, video, or computer-based). This can be as humble as a typical cassette tape recorder to as lavish as equipment found in sophisticated testing labs—such recording equipment might include several video cameras; an eye tracking camera; audio, dubbing, mixing equipment; a time stamping program; as well as keystroke-tracking programs.
3. A place to conduct the protocol. Depending on writers' goals, they will want to test the text in either its actual environment (e.g., on the plant floor or in a busy office) or under quiet, laboratory-like conditions.
4. The equipment described in the text if collecting a user protocol. That is, any other equipment needed to conduct the protocol, depending on the text writers are testing (e.g., if the text is a set of instructions for using cuisinart, writers will need a cuisinart).
5. Two copies of the text to be evaluated (one for the writer and one for the participant).
6. Tapes (either audio or video). Writers should be certain to have a backup tape. They should test the tape(s) before collecting the protocol to be sure the recorder is functioning properly and that the tape is recording well enough so that one can distinguish words that are spoken in a soft voice.
7. A place to observe. The writer collecting the protocol should have a place to observe, preferably a place that will not make the participant feel uncomfortable. Many writers prefer to sit behind a two-way mirror. If it is necessary to sit next to the participant, the chair should be positioned slightly behind the participant; thus, the writer can look at the screen or the text the participant is reading, and make the participant feel less like he or she is being observed.

In addition, writers may want to devise a preliminary coding scheme. Coding protocols proceeds much more efficiently when the writer has at least some idea of the kinds of difficulties the document may produce. Before collecting the protocol, it is important to have already conducted a technical and stylistic review. Such reviews allow writers to begin protocol-aided revision with the best draft possible, that is, one that has been checked for technical accuracy and style. In addition, these reviews often become good sources of ideas for a preliminary coding scheme, for example, errors caused by descriptions, errors caused by procedures, errors caused by poor formatting, or errors caused by missing information.

Collecting a protocol

Before collecting a protocol, writers will need to consider the practical issues mentioned above. In addition to providing participants with a well-written set of instructions, it is wise to play an audio or video demonstration tape of someone giving a protocol. The demonstration tape should be about two minutes in length and should illustrate a range of positive and negative comments about the text, for example, signs of approval, questions, confusions, predictions, elaborations, or any reading behavior. The goal is to provide participants with an example that shows them *not* "what to do or what is expected," but the range of ways people respond to texts. The aim is to make participants feel comfortable in responding with whatever comes to their mind as they are engaged in the task. Clear instructions along with a sample tape will typically be enough to make most participants feel at ease about talking aloud as they read and/or perform a task.

As the participant is providing the protocol, writers should try to catalog all they see, including nonverbal behavior. They ought to follow along as the participant reads the text so they can mark any section that is unclear or confusing. Once a protocol is in progress, it is best if the writer does not intervene. If participants have questions, writers should allow them time to figure out the answer on their own. Writers must resist intervening unless the participant becomes frustrated and wants to stop the protocol. After the participant has completed the task, writers should answer questions, thank him or her, and explain the project in more detail.

Fixing problems the text creates for participants

The last and most important stage in protocol-aided revision is fixing the problems created by the text. When the text causes few problems, writers can usually solve its problems by making deletions and minor additions. But when readers are confused by many aspects of the text, major revisions and rewriting are often necessary. It is important for writers to try to determine the locus of the problems. In this way, they will have better information about what solution strategies to employ. Sometimes readers will misunderstand an entire text, but the misunderstanding may arise out of one fundamental misconception that occurs early in the text. Other times, the problems will be distributed throughout the text and the reader's difficulties may escalate with almost every new idea.

Once the protocol has been coded, writer will have a better sense of which problems are most severe and/or frequent. Since most revision is done while under time pressure to finish, writers must attend to the most severe problems and should decide which problems, if any, will be left unsolved. The goal, of course, is to find solutions to as many problems as time permits.

Choosing optimal revision strategies is also a skill that develops with extensive practice. While there are no clear-cut strategies that work in every revision situation, a few key questions should initiate any revision activity:

- "What problems are created by the organization, structure, or layout of the text?" Answering this question will help focus the writer's attention on the text's macrostructure. Research in text design shows that global features such as structure and levels of subordination, e.g., headings and subheadings, have the most impact on the memorability of a text (Britton 1986). Similarly, research in writing underscores that skilled revisers attend to the text's global features early in their process, and that solving high-level problems first often has the effect of eliminating lower-level problems at the same time (Hayes et al. 1987; Schriver 1989b).
- "What alternative verbal or visual solutions are available?" Sometimes, as is demonstrated in the case study that follows, a visual solution can help to solve most of the text's problems. For functional documents, visual solutions usually take one of two shapes. One way lies in changing the typography, section headings, margins, rules, or layout. Another more obvious means of creating a visual solution is to supplement or substitute the text with pictures, diagrams, tables, charts, or other graphic devices.

The goal in asking these questions is to solve the text's most severe problems in the shortest amount of time. Research shows that writers who revise by adopting a sentence-level perspective of the text typically fail to make revisions that increase the effectiveness of the whole text (Sommers 1980; Hayes et al. 1987; Schriver 1987, 1988, 1989b). A sentence-level perspective is one in which revisers attack text problems linearly. They begin by reading the first sentence of the text and by asking "Is there anything wrong with this sentence? If they find a problem, they fix it and proceed to the next sentence. The drawback of this strategy is that it blinds revisers to how the whole text is functioning and focuses their attention to word and sentence-level errors (Schriver, 1987, 1988). While it is essential to fix local errors, it is more important and more efficient to adopt a whole-text perspective. In this way, the most pervasive problems will be dealt with first, thus allowing the solution of sentence-level problems without the possibility of wasting time fixing problems locally and then determining that the whole text needs to be rewritten.

PROTOCOL-AIDED REVISION IN PRACTICE: A CASE STUDY OF THE PROCESS

To help writers see how protocols can be used in revision, the following case study illustrates the process of protocol-aided revision. The original text on which this case study is based, "The DUC System," comes from the beginning of a tutorial section of a computer manual on the topic of computer-aided design. The tutorial is aimed at graphic designers who are experienced in design using pen and paper but who have no familiarity with design using a computer. Its goal is to teach new users the fundamentals of computer-aided design. The manual was written to accompany a new Design Using Computers (DUC) system, one of the early computer-aided design (CAD) systems.

In 1983, this particular tutorial was being used by a CAD lab within the design department at Carnegie Mellon University to teach undergraduate graphic designers the basics of CAD systems. The director of the CAD lab asked the document design team at the Communications Design Center to revise the text because he felt the text took students too long to "get started" and that they were making errors in using the equipment. To determine the nature of the text's problems for students, writers chose to collect user protocols to evaluate the manner in which the tutorial was being read and used. Participants who volunteered to provide user protocols were senior undergraduate design majors who had no prior experience in using CAD equipment.

This case study is intended to illustrate how protocols can help writers see the need to substitute or supplement their verbal text with *visual text*. Originally, it appeared as one of ten lessons in revising computer documentation for comprehension using protocol-aided revision (Schriver 1984). It is organized into three parts:

- the original text, "The DUC System" (figure 1),
- a sample user protocol from one of the inexperienced "DUC" users (figure 2), and
- a protocol-aided revision based on the inexperienced user's difficulties with the text (figs. 3a–d).

Figure 2, the user's protocol, shows the variety of problems the design student had with the text. The problems he experienced while trying to use the tutorial allowed writer Carol Janik to diagnose problems of several types. First, problems that took the shape of "what" questions, signaling a call for definitions and purpose statements. For example, "What is the purpose of the light pen?" "What is the difference between selecting and indicating?" "What does TC stand for?"

Second, problems that took the shape of "how" questions, signaling the need for better procedural information. For example, "How do I hold the light pen?" "Do I simultaneously hit both indicating and function keys?" "Couldn't they give me a better idea of how to hold the light pen with a drawing instead of words?" "Is this really a two-step action, first you select and then you indicate?"

Third, problems that took the shape of "where" questions, signaling the need to clarify the location of various areas of the screen as well as where the user should look in order to get feedback and/or confirm that his actions were accurate. For example, "Where should I be looking to get the point of this information?" "Will the terminal always prompt me for these commands?" "Where exactly is the message versus the menu area?" "Where is the indicate function key located?"

The Display Station

Each display station consists of a terminal with a display screen, a typewriter keyboard, a light pen, and a function keyboard.

The Display Screen

Message Area—provides feedback on the current status, e.g., function currently in use, scale of drawing, warning of invalid operation, etc.

Menu Area—provides options which you can choose with the light pen.

The Light Pen

The light pen is a device which serves two functions. You will learn how to draw points to lines to more complex objects such as circles and ellipses.

Selecting (Sel)

Selecting an item on the menu area or an element in the drawing area. Hold the pen perpendicular to the screen with the point touching the desired item and then push the pen point into the screen until the pen clicks. The terminal prompts you to select by typing Sel in the message area of the display screen.

Indicating (Ind or Tc)

Indicating an item on the drawing area of the screen. Hold the pen perpendicular to the screen at approximately the desired place and then hit the Indicate function key. The terminal will prompt you to indicate by typing IND or TC in the message area of the display screen.

Both of these functions will be covered in the first exercise.

Throughout the manual, we will designate instructions as follows:

- "Select" always applies to pushing the light pen into the screen until the pen clicks.
- "Indicate" always applies to touching the light pen to the screen at the specified place and hitting the Indicate function.

Figure 1. The original text, "The DUC System."

Well, if I wasn't sitting in front of this terminal, I'd never guess this stuff is about a computer-aided design system.

The Display Station. Each display station consists of a terminal with a display screen, a typewriter keyboard, a light pen, and a function keyboard. *OK, so I'm looking at this equipment and trying to tell which is which. I've never used a function box or a light pen before. I want to try out this light pen; it looks fun to use. The only thing is that I'm left-handed and I don't know how to position my hand with all this new stuff.*

The Display Screen. *Well, I'll assume this means the terminal screen.* Message Area—provides feedback on the current status, e.g., function currently in use, scale of drawing, warning of invalid operation, etc. *I really don't know what I should be looking at here. What part is the display screen. I haven't tried anything yet so I don't know why they're telling me this detail now. I don't yet understand current status or invalid operations either.* Menu Area—provides options which you can choose with the light pen. *Alright, but where exactly in the menu area?*

The Light Pen. *What is the purpose of the light pen?* The light pen is a device which serves two functions. You will learn how to draw points to lines to more complex objects such as circles and ellipses. *Huh? OK, I'm trying to learn the function of this equipment all at once because I'm eager to create my own drawings. I guess for this session, I'll have to be satisfied with the basics. From this introduction, I expect that I'll be moving from drawing points to lines to more complex elements like circles and ellipses. OK. I'm ready. I wonder which functions they are referring to—drawing lines and circles or selecting and indicating?*

Selecting (Sel). Selecting an item on the menu area or an element in the drawing area. Hold the pen perpendicular to the screen with the point touching the desired item and then push the pen point into the screen until the pen clicks. *I know it says what to do, but I'm having difficulty knowing exactly how to hold this thing. How do you make it perpendicular? I don't know how hard to press and I'm afraid of damaging the screen.* The terminal prompts you to select by typing Sel in the message area of the display screen. *I wonder if it always prompts you?*

Indicating (Ind or Tc). Indicating an item on the drawing area of the screen. Hold the pen perpendicular to the screen at approximately the desired place and then hit the Indicate function key. *Where is the indicate function key located? How can I do both indicating and hitting function keys simultaneously?* The terminal will prompt you to indicate by typing IND or TC in the message area of the display screen. *What does TC stand for? Where is the message area of the screen? Couldn't they give me a better idea of how to hold the light pen with a drawing instead of words?*

Both of these functions will be covered in the first exercise. *What is the essential difference between selecting and indicating? I'm not getting this from this description. Why are they telling me these commands without a context?*

Throughout the manual, we will designate instructions as follows. *Is there a reason to tell me the instructions without telling me when and where I will use them?*

"Select" always applies to pushing the light pen into the screen until the pen clicks. *Is this really a two-step action? First you select and then you indicate? Or are there times when I simply indicate?*

"Indicate" always applies to touching the light pen to the screen at the specified place and hitting the Indicate function. *Sounds fairly clear, but I still don't know in what contexts I'd choose these. It seems odd to put them here.*

Figure 2. A user protocol collected from an inexperienced user of "The DUC System."

The Display Station

Each display station consists of a terminal with a display screen, a function keyboard, a typewriter keyboard, and a light pen. If you are right handed, it is usually best to hold the light pen in your right hand and position the function keyboard on your left. If you are left handed, do the opposite.

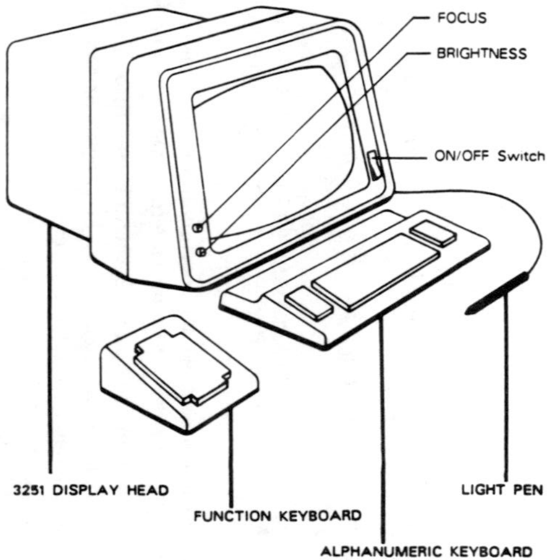

The Light Pen

The light pen serves two functions:

- **selecting** a graphic element (such as a point, line or circle) on the screen or an item on the menu (a list of options on the screen)
- **indicating** an element, an area, or a direction on the screen

You will learn how to use the light pen in both of these ways on pages 23 and 39 of Exercise One.

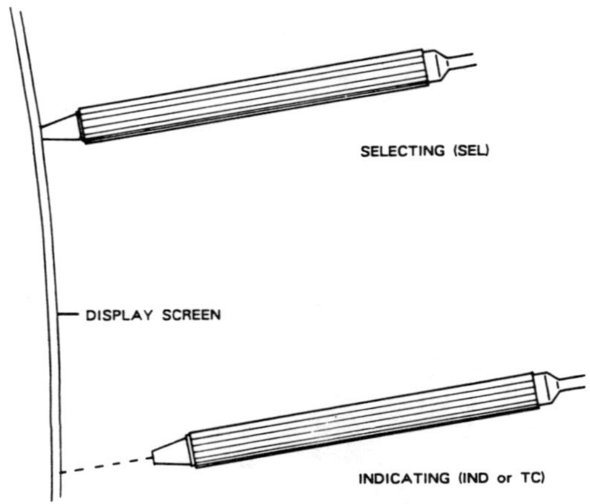

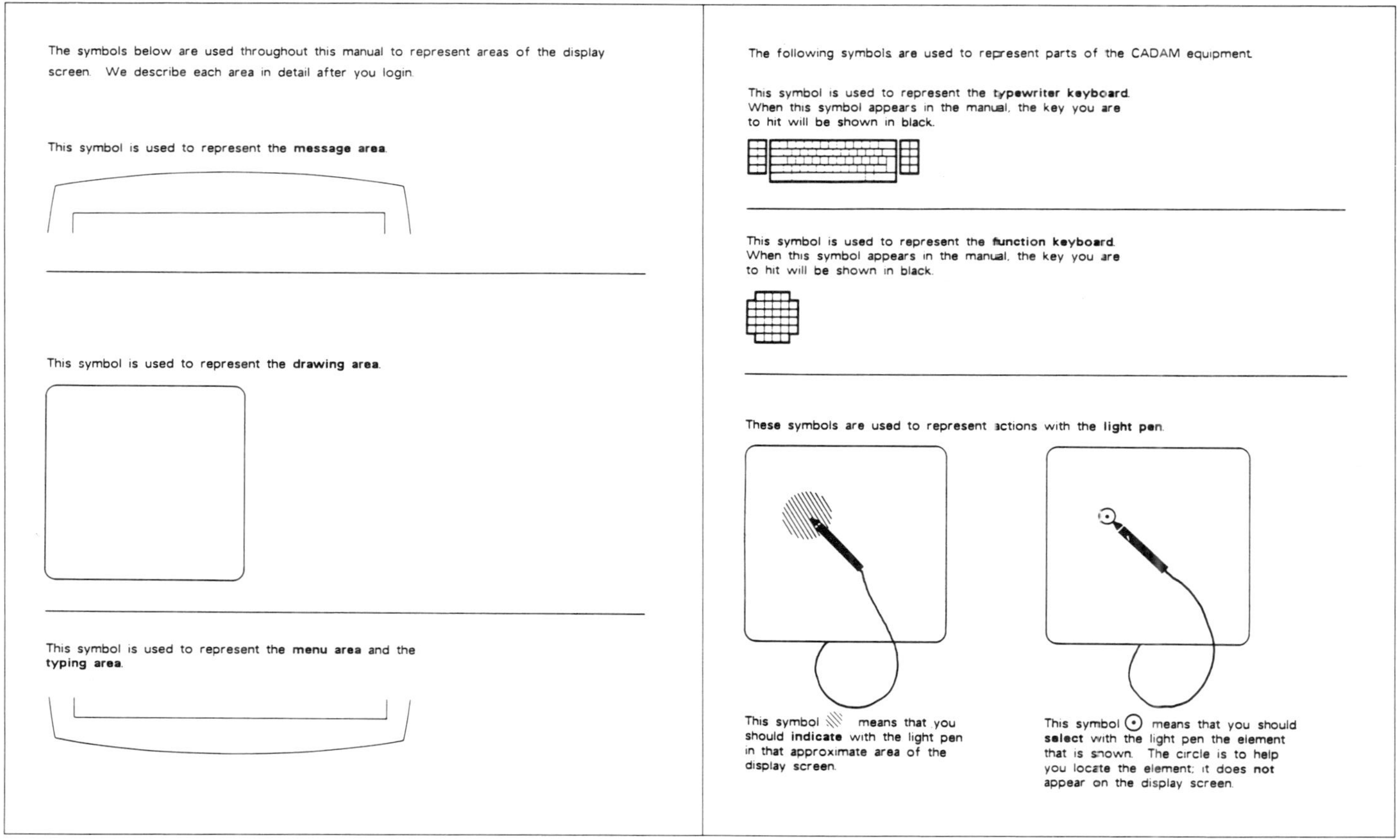

The symbols below are used throughout this manual to represent areas of the display screen. We describe each area in detail after you login.

This symbol is used to represent the **message area**.

This symbol is used to represent the **drawing area**.

This symbol is used to represent the **menu area** and the **typing area**.

The following symbols are used to represent parts of the CADAM equipment.

This symbol is used to represent the **typewriter keyboard**. When this symbol appears in the manual, the key you are to hit will be shown in black.

This symbol is used to represent the **function keyboard**. When this symbol appears in the manual, the key you are to hit will be shown in black.

These symbols are used to represent actions with the **light pen**.

This symbol means that you should **indicate** with the light pen in that approximate area of the display screen.

This symbol means that you should **select** with the light pen the element that is shown. The circle is to help you locate the element; it does **not** appear on the display screen.

Figure 3. A protocol-aided revision for an inexperienced user of "The DUC System."

Fourth, problems that took the shape of "why" questions, signaling a call to provide more contextual information about the user's goals in invoking particular commands. For example, "Why are they telling me these commands without a context?" "Is there a reason to tell me the instructions without telling me when and where I will use them?"

In deciding how to revise the tutorial, Janik felt that the user's problems were too numerous and too severe to warrant solving the text's problems by making minor sentence-level repairs. She concluded that the user's comments suggested that the primary difficulty with the text was that it relied exclusively on a verbal presentation when a graphic presentation was needed. Thus, she tried to solve the user's questions with one major strategy, that is, changing the text from a verbal to visual presentation.

Janik rewrote the text (figure 3), starting by creating a diagram designed to answer the user's question about the equipment and "how to hold the light pen." The revision eliminates most of the reader's problems by including a drawing of how to position the light pen in relation to the display screen. The revised information on the light pen specifies its basic functions and then previews that readers will learn to use the pen later in the first exercise. Since users need to know information about other system functions and procedures, such as logging on and using the function keyboard *before* they can actually begin to draw, she separated the information accordingly. She also adopted graphic conventions for the meanings of *selecting* and *indicating*—conventions that were used throughout the manual.

Janik's primary goal during revision was to make the text's organization more transparent. To do so, she reorganized the text, changed the page layout, and divided the information into manageable chunks for the reader. She also aimed to provide the reader with a better sense of the consequences of specific actions. The revision tells the reader specifically that *selecting* relates to points, lines, or circles while *indicating* is used for elements, areas, or directions on the screen.

A PROCESS MODEL OF PROTOCOL-AIDED REVISION

As a way to help writers use protocols to guide revision, figure 4 presents a process model of protocol-aided revision. The model is intended to help writers see the relationship among three components: (1) cognitive processes, (2) writer and reader activities, and (3) outputs of processes and activities. The first column, cognitive processes in protocol-aided revision, is influenced by the revision model designed by Hayes, Flower, Schriver, Stratman, and Carey (1987). This model is designed to capture the cognitive processes involved in using protocols to revise. Protocol-aided revision, like other sorts of text revision, involves four key subprocesses:

1. *Task Representation*—representing the text's goals, constraints, and criteria for success;
2. *Detection*—seeing or noticing problems;

Cognitive Processes in Revision	Writer/Document Designer Activities	Output of Processes and Activities
Represent Task	• establishes goals; • analyzes audience; • predicts audience problems; • defines constraints; • establishes criteria for success.	• a plan for approaching the revision which considers both goals and constraints; • an analysis of audience needs and potential difficulties with the text; • a set of criteria for evaluating the text's success.
Detect Problems	• reads, listens to, or views each protocol with the goal of noting readers' difficulties; • attends to problems caused by errors of commission and/or omission.	• a set of comments and/or behaviors which can be viewed as flagging possible areas/aspects in need of revision.
Consolidate Reader Comments	• identifies comments that signal problems and tabulates their frequency; • identifies comments that are positive evaluations; • identifies the location of problems on the source text.	• a list of negative comments and their frequency; • a list of positive comments; • a source text on which the location of problems are identified.
Identify Important Problems	• evaluates comments for their relevance to the goals and criteria for the text's success; • chooses problems to fix on the basis of their frequency and importance.	• a prioritized list of important problems that need to be solved.
Diagnose Problems	• characterizes the nature of readers' problems; • classifies the important problems by type.	• a classified set of problem types that may suggest effective and/or efficient revision strategies.
Select Strategies	• considers optional means for solving, e.g., revise locally, paraphrase, or rewrite; • considers possible visual and verbal solutions to problems; • chooses revision strategies influenced by type and importance of problems; • uses readers' positive comments to aid in revising.	• a range of options for creating global and local text revisions, e.g., add, delete, substitute or modify; • a set of alternative visual and/or verbal solutions to the text's problems; • ideas for using readers' positive comments as aids to revision.
Fix Problems	• fixes problems by repairing, modifying, rewriting, or reconceptualizing the text.	• a revised text ready for comparison against goals/criteria for success.

Figure 4. A process model of protocol-aided revision.

3. *Diagnosis*—characterizing or describing what the problem is;
4. *Strategy Selection*—choosing methods for solving identified problems.

Within each of these subprocesses, writers have a wide variety of options. The ability to exercise these options and the ability to choose and carry out effective revisions has been shown to distinguish experienced from inexperienced writers.

A *task representation* is a set of goals and criteria which the writer uses to guide the revision process and to evaluate the final product. The writer represents the task by considering issues such as: the document's goals (such as, to instruct or persuade); the audience's needs (for example, to use the text to learn a new procedure or to make a decision based on the text's content); the audience's potential problems with the text (for instance, they may fail to understand the procedures or could misinterpret the content of the text); and the inevitable constraints under which revision must take place (typically lack of time and resources).

The second subprocess, *detection,* or seeing problems in a text, is a fundamental revision skill because if the writer never sees problems in the first place, nothing gets revised. Writing instructors agree that teaching detection skills is very difficult, and that better ways are needed to help writers see problems in text.

Detection is a skill that seems to vary depending on authorship and knowledge. Research shows that difficulty in detecting problems in texts depends, in part, on whether writers wrote the text themselves or whether it was created by a writer different from the reviser. Writers typically have more difficulty seeing problems in their own text than in those created by someone else (Hull 1984). Writers are often too close to the intended meaning in their text to see it as representing anything less than their intentions. They often view their text as communicating more effectively than it actually does for the intended reader. Consequently, authorship of the text, that is, whether the writers are revising their own or someone else's text, is an important factor determining success in revision.

Another barrier to the success of the detection process is topic knowledge. Research shows that writers with substantial topic knowledge of the text's main ideas often have significant trouble detecting problems that their texts create for readers without such knowledge (Bond et al. 1980; Hayes et al. 1986; Schriver 1987, 1989a). In "If it's Clear to Me, it Must be Clear to Them," Hayes, Schriver, Blaustein, and Spilka (1986) describe "the knowledge effect" in writing and how topic knowledge prevents writers from seeing problems in text. Topic knowledge was found to act as a "blinder" to text problems. High-knowledge writers tended to overestimate their audience and believed that what was understandable to them would be clear to anyone.

Similarly, in a study of teaching writers to predict readers' needs, I found that upper-class undergraduate writers with two years of experience with word processing were extremely insensitive in their ability to predict problems that freshman users would have with poorly written word processing manuals (1987). The knowledge effect, then, may be the unseen culprit behind why "high-knowledge experts" such as lawyers, doctors, computer scientists, engineers, economists, and government representatives frequently produce incomprehensible texts. It may also provide a clue as to why so many university professors have difficulty in communicating "the basics" to freshmen in introductory college courses.

An implication from the research in detection is that writers who are revising their own text and who have high topic knowledge may be at a considerable disadvantage in seeing the problems the text may create for readers. One of the most important cognitive advantages of protocol-aided revision is that it provides a method for "getting around" the effects of knowledge and authorship. *Unlike standard revision procedures which place the responsibility of detecting problems on the writer, protocol-aided revision places the burden of detection on the reader.* Detection of text problems is carried out while readers are engaged in trying to understand and/or use the document. The reader's comments, questions, rereadings, and hesitations can be viewed as flagging possible problems. Very often, the protocols will help writers detect both problems of commission, that is, problems caused by what the text says, and problems of omission, that is, problems caused by what the text is missing (Schriver 1987).

Once possible problem areas have been detected, writers are faced with the somewhat mundane yet necessary evaluative process of *consolidating readers' comments* across all protocols. Consolidation of readers' comments should be done in two complementary ways. The first involves simply identifying the location of text areas where readers experience problems. The second involves using the individual protocol transcripts to create a master list of problems across all participants. This list should summarize all candidate problems for revision and tabulate their frequency.

Along with listing what readers disliked or had problems with, it is a useful strategy to list those aspects of the text that readers liked and had no problems with. While protocols mainly provide information related to comprehension and use difficulties, readers sometimes comment on what they like about the text. When this occurs, writers should ask themselves, "What am I doing right?" "Can this successful part be repeated or done better?" Quite often the successful parts of the text can be amplified or used as a model for less successful parts.

The next process in protocol-aided revision involves *identifying the important problems* to attend to. Given the typical constraints under which writers revise, they must give priority to a subset of the text's problems. In identifying important problems, writers isolate those problems which most inhibit the text's success.

After problems to be revised are determined, *diagnosis,* the process of characterizing the nature of the problem, becomes important. In diagnosis, writers must answer "What is the cause of the reader's problem?" Isolating the origin of the problem is usually much of the work in finding its solution (Flower, Hayes, Carey, Schriver, and Stratman 1986). Very often, identifying important problems and diagnosing problems occur simultaneously; because the writer can identify the cause of the problem, he or she can judge its importance relative to the whole text.

Strategy selection is the act of considering optional means for solving the text's problems. Quite often, the process of diagnosing the text's problems suggests effective revision strategies. In selecting strategies, writers are influenced by the type and importance of problems. Some problems, such as lack of coherence, simply warrant more effort than others. In addition, writers are influenced by the constraints under which revision takes place. Very often, factors such as time and cost exert a major influence on our final decisions for revision. The impact of constraints on the selection of revision strategies is an important and unexplored area in writing and document design research (Schriver 1989a). It is clear that writers need better advice on how to make design decisions under severe constraints.

During strategy selection, writers aim to identify visual and/or verbal solutions to text problems. For any solution that involves both visual and verbal text, they must also consider how to best integrate their proposed solutions. The output of strategy selection is a representation of ways to solve the text's problems—revise locally, paraphrase, rewrite, or reconceptualize.

After decisions have been made about what to do, writers can then try to *fix the text's problems* through activities such as repairing, modifying, or rewriting the text. Once the writer finishes a complete cycle of protocol-aided revision, the text should be compared against its goals and criteria for success. The results of this comparison will tell the writer if another cycle of protocol-aided revision is needed.

Overall, as the model suggests, protocol-aided revision is a recursive process which asks the writer to use reader feedback as a basis for guiding (and ending) revision. For a more exhaustive discussion of the cognitive processes involved in protocol-aided revision, see Schriver (in press).

IMPLICATIONS OF USING PROTOCOL-AIDED REVISION FOR WRITERS

There is now empirical evidence which shows that writers who use protocol-aided revision can improve their skills in anticipating readers' problems with poorly written text. Recent writing research shows that writers who use protocols to revise acquire new abilities in predicting readers' responses to poorly written text—abilities which transfer to writing contexts in which protocols are unavailable. In a study of teaching writers to anticipate the reader's needs, I (1987) found that writers taught to detect and diagnose readers' problems with the help of think-aloud protocols improve significantly in their ability to *predict readers' problems in texts where no protocol is available*. This improvement was not found in writers taught "to consider the audience" using standard methods of audience analysis, for example, peer critiquing methods and demographic heuristic techniques.

Writers taught with protocols showed dramatic improvement in their ability to predict readers' problems (a 62% increase in accurate predictions) after their careful analysis of a sequence of ten lessons in protocol-aided revision. Each lesson consisted of a poorly written text (that is, one that will give the intended audience difficulties in understanding it), and a think-aloud protocol of a person trying to understand the text (Schriver 1984). For each of the ten lessons, writers were asked to predict the intended audience's problems with the poorly written text by employing a two-step procedure. First, writers read the poorly written text with the goal of detecting and diagnosing readers' potential problems with the text. Second, writers evaluated the think-aloud with the goal of detecting and diagnosing additional problems that were made evident in the protocol—problems that writers missed on their first pass. Writers repeated this procedure for each of the ten lessons. To assess the effectiveness of the teaching, writers taught with protocol-aided revision (the experimental group) and writers taught in "audience analysis" (the control group) were given a pretest and a post-test which evaluated their ability to detect and diagnose problems with poorly written texts on which protocols were not available.

Writers taught to anticipate readers' needs by evaluating protocols improved significantly in their ability to detect and diagnose problems caused by incomplete information—problems such as missing contexts, missing purpose statements, missing examples, and missing procedures. This research provides strong support for the important role that reader feedback can play in improving writers' detection and diagnosis skills. It also suggests that writers with practice in evaluating protocols are more likely to produce better first drafts because they are better able to anticipate a reader's response to the text.

Moreover, this paper points out that early conceptions of plain language were unsuccessful in providing writers with concrete ways for revising texts to meet the unique needs of particular audiences. Furthermore, as the case study makes apparent, writers' solutions to communication problems are sometimes constrained by their neglect of the visual expression of ideas. The goal here was to extend conceptions of plain language and to suggest that protocol-aided revision can help writers achieve their rhetorical goals for readers in several important ways:

1. It can help writers detect and diagnose the difficulties created by what the text says or by what it fails to say—difficulties that can inhibit intelligibility and usability.
2. It can help writers identify the need for creating visual solutions to text problems—typography, graphs, pictures, format, diagrams, flowcharts, tables, and so on—and for integrating their visual and verbal decisions.

Methods such as protocol-aided revision that focus on helping writers become more sensitive to the complex needs of their intended readers deserve careful attention and further research.

REFERENCES

American Institute for Research. *Simply Stated* (formerly *Fine Print*). Washington, DC: American Institute for Research, Document Design Center.

Atlas, M. A. "The user edit: Making manuals easier to use." *Institute of Electrical and Electronics Engineers (IEEE) Transactions on Professional Communication,* PC-24 (1981): 28–29.

Berkenkotter, C. "Decisions and revisions: The planning strategies of a publishing writer." *College Composition and Communication,* 34, no. 2 (1983): 156–69.

Bond, S. J. "Protocol-aided revision: A tool for making documents usable." Paper presented at the IBM Academic Information Systems University AEP Conference, Alexandria, VA, 1985.

Bond, S. J.; Hayes, J. R.; and Flower, L. S. "Translating the law into common language: A protocol study." Document Design Project Technical Report 8, Carnegie Mellon University, Pittsburgh, PA, April, 1980.

Britton, B. "Capturing art to improve text quality." *Educational Psychologist,* 21, no. 4 (1986): 333–56.

Charrow, V. "What is plain English, anyway?" Document Design Center, Technical Report, Washington, DC: American Institutes for Research, December, 1979.

Communications Design Center. *Communications Design Center Newsletter.* Pittsburgh, PA: Carnegie Mellon University.

Duffy, T. M. "Readability formulas: What's the use." Duffy, T. M., and Waller, R. *Designing usable texts.* Orlando, FL: Academic Press Inc., 1985.

Ericsson, K. A., and Simon, H. A. *Protocol analysis: Verbal reports as data.* Cambridge, MA: MIT Press, 1984.

Felker, D. B. *Document design: A review of the relevant research.* Washington, DC: American Institute for Research, 1980.

Flesch, R.' *The art of readable writing.* 1949. New York, NY: Harper and Row Brothers Publishers, 1974.

Flower, L.; Hayes, J. R.; Carey, L.; Schriver, K. A.; and Stratman, J. "Detection, diagnosis and the strategies of revision." *College Composition and Communication,* 37, no. 1 (1986): 16–55.

Gunning, R. *The technique of clear writing.* 1952. New York, NY: McGraw-Hill Book Company, 1968.

Hayes, J. R., and Flower, L. S. "Identifying the organization of writing processes." *Cognitive processes in writing.* Eds. L. W. Gregg and E. R. Steinberg. Hillsdale, NJ: Lawrence Erlbaum Associates, 1980.

Hayes, J. R., and Flower, L. S. "Uncovering cognitive processes in writing: An introduction to protocol analysis." *Research in writing: Principles and methods.* Eds. P. Mosenthal, L. Tamor, and S. Walmsey. New York, NY: Longman, 1983.

Hayes, J. R.; Schriver, K. A; Blaustein, A.; and Spilka, R. "Is this text clear?—How knowledge makes it difficult to judge." Paper presented at the American Educational Research Association (AERA Conference), San Francisco, CA, April, 1986.

Hayes, J. R.; Flower, L.; Schriver, K. A.; Stratman, J.; and Carey, L. "Cognitive processes in revision." *Advances in Applied Psycholinguistics, Volume II: Reading, Writing, and Language Processing.* Ed. S. Rosenberg. Cambridge, England: Cambridge University Press, 1987. 176–240.

Hull, G. "A performance study of the editing process in writing." Doctoral dissertation, Pittsburgh, PA: University of Pittsburgh, Department of English, 1984.

Information Design Journal. Milton Keynes, England: The Open University.

Law Reform Commission of Victoria. "Legislation, legal rights and plain English." Discussion Paper No. 1. Law Reform Commission of Victoria, 7th Floor, 160 Queen Street, Melbourne, Australia, 1986.

Lewis, C., and Mack, R., "Learning to use a text processing system: Evidence from 'thinking-aloud' protocols." *CHI '82 Human Factors in Computing Systems Conference Proceedings.* New York: Association for Computing Machinery, 1982: 387–92.

Lund, M. A. "Evaluating the user interface: A candid camera approach." *Computer and Human Interaction (CHI) '85 Proceedings.* New York: Association for Computing Machinery, 1985, 107–13.

MacNeil/Lehrer Report. "Plain English." Library #625, Show 3165. WNET/WETA, New York, NY: Educational Broadcasting Company, February, 1978.

Payne, J. W., and Ragsdale. "Verbal protocols and direct observation of supermarket shopping behavior: Some findings and a discussion of methods." *Advances in Consumer Research,* Vol. 5. Ed. H. K. Hunt. Chicago, IL: Association for Consumer Research, 1978.

Roberts, D. D., and Sullivan, P. A. "Beyond the static audience construct: Reading protocols in the technical writing class." *Journal of Technical Writing and Communication* 14 (1984): 143–53.

Schriver, K. A. "Revising Computer Documentation for Comprehension: Ten exercises in protocol-aided revision." Technical Report No. 14. Pittsburgh, PA: Carnegie Mellon University, Communications Design Center, 1984.

———. "Teaching writers to anticipate the reader's needs: Empirically-based instruction." Doctoral dissertation, Carnegie Mellon University, Department of English, Pittsburgh, PA, 1987.

———. "Moving from sentence-level to whole-text revision: Helping writers focus on the reader's needs." In *Expanding the Repertoire,* ed., Kathy McCormick, Center for the Study of Writing, Technical Report No. 30. Berkeley, CA and Pittsburgh, CA (1988): 46–57.

———. "Document design from 1980 to 1989: Challenges that remain." *Technical Communication,* 4th Quarter (1989): 316–31.

———. "Evaluating text quality: The continuum from text-focused to reader-focused methods." *IEEE Transactions on Professional Communication* 32, no. 4 (Dec. 1989b,): 238–55.

———. "Plain language for expert or lay audiences: Designing text using protocol-aided revision." Center for the Study of Writing Technical Report. Berkeley, CA, and Pittsburgh, PA: Univ. of California at Berkeley and Carnegie Mellon (in press).

Schriver, K. A.; Hayes, J. R.; Danley, C.; Wulff, W.; Davies, L.; Cerroni, K.; Graham, D.; Flood, E.; and Bond, E. *Designing computer documentation: A review of the literature—hardcopy, online, general applications.* Technical Report No. 31. Pittsburgh, PA: Carnegie Mellon University, Communications Design Center, 1986.

Simon, H.A., and Gilmartin, K.A. "A simulation of memory for chess positions." *Cognitive Psychology* 5 (1973): 29–46.

Smargorinsky, P. "The reliability and validity of protocol analysis." *Written Communication* 6, no. 4 (1989): 463–79.

Soderston, C. "The usability edit: A new level." *Technical Communication,* 1st Quarter (1985) 16–18.

Sommers, N. I. "Revision strategies of student writers and experienced writers." *College Composition and Communication* 31 (1980): 378–87.

Strunk, W. Jr., and White, E. B. *The elements of style.* New York: Macmillan, 1959.

Sullivan, P., and Flower, L. "How do users read computer manuals? Some protocol contributions to writers' knowledge." *Convergences: Transactions in reading and writing.* Ed. B. T. Petersen. Urbana, IL: National Council of Teachers of English, 1986. 163–78.

Swaney, J. H.; Janik, C. J.; Bond, S. J.; and Hayes, J. R. "Editing for comprehension: Improving the process through reading protocols." Document Design Project Report No. 14, Pittsburgh, PA: Carnegie Mellon University, 1981.

Visible Language. 2643 Easton Road, Cleveland, OH, 44118.

Winbush, B., and McDowell, G. "Testing: How to increase the usability of computer manuals." *Technical Communication*, 4th Quarter (1980): 20–22.

Winkler, Ferguson, and Youngquist. "Procedures for designing and evaluating training manuals." *Proceedings of the Society for Technical Communications.* Washington, DC: Society of Technical Communications, 1985, RET-6-9.

Witman, W. "Using reading protocols to edit documents: A case study." *Proceedings of the Society for Technical Communications.* Washington, DC: Society of Technical Communication, 1987, RET-159-161.

Editing for Comprehension: Improving the Process through Reading Protocols

Joyce Hannah Swaney, Carol J. Janik, Sandra J. Bond, and John R. Hayes

This series of experiments demonstrates that (1) standard editing techniques, even when applied by skilled document designers, sometimes fail to improve the clarity of a document, and (2) reading protocols provide a powerful editing tool which can yield improvement in clarity where standard techniques fail.

In the first experiment, four documents were revised for clarity using standard editing techniques. Comprehension tests revealed that while three of the documents were improved by revision, the fourth appeared to have been made worse. In the second study, protocols of subjects reading the fourth document were collected to pinpoint comprehension difficulties. The document was then revised to reduce these difficulties. Comprehension tests revealed that the error rate for the final revision had dropped dramatically.

Editing for clarity is of central importance in document design. However, the literature suggests that even in skilled hands, standard editing procedures[1] are limited in their effectiveness. Skilled editing often improves clarity, but in a disturbing minority of cases it fails to do so and sometimes even makes things worse.

The purpose of this study is to explore the process of skilled editing, in order to understand why it may fail and what can be done to make it more effective.

OTHER RESEARCH

Most of the experiments on comprehension have been done at the sentence level, but clearly it is also important that experiments be done using whole texts. Sentences are rarely used in isolation; all the parts of a text must work together. In studies using whole documents, revisions which were designed to improve comprehensibility often led to less than satisfactory results.

In some cases a revision resulted in better overall comprehension, but less comprehension on particular items. Charrow, Holland, Peck, and Shelton (1980), in a study of a revised Medicaid Recertification form, found that although people generally understood the revised form better, they still found it difficult and on a few items they actually performed worse with the revision. Jeffrey Davis (1977) of the New York University Law School tested a revised consumer credit contract on 148 adults, asking them fifteen multiple-choice questions. Overall, the revision showed an improvement of 26%. Those who benefited most from having the revision were young, had a low income, or were inexperienced with credit. But, again, on some questions more errors were made with the revised contract than with the original.

Sometimes revisions improved comprehension, but only for selected groups of users. For example, in the Davis study mentioned above, the group that one would expect to have the most trouble with the document benefited most from the revision. However, this was not the case in a study by Kincaid, Kincaid, and Lewis (1977). They tested a revised installment loan form on two groups of subjects. One group was composed of college students who had been judged to be average readers, and the other group was night college students who were judged to be above-average readers. About 70% of those in the latter group were considerably older than those in the first group, and nearly half had had at least one installment loan. Kincaid et al. used the cloze test to measure comprehension.[2] The above-average readers showed a significantly better performance with the revision. The average readers did not show a significant improvement; in fact their scores on both versions were below the minimum score that Kincaid et al. believed was necessary to indicate comprehension.

Despite these mixed findings, the revisions in the studies cited did result in an overall improvement in understanding, at least for some people.

EXPERIMENT 1: REVISING FOUR DOCUMENTS

In experiment 1 four documents were revised using standard editing techniques. Then both the originals and the revisions were tested on people chosen from the general public. Our goal was to determine if the revisions improved comprehension.

The editing was done jointly by four document design experts, each with considerable experience in writing and editing. They were very aware of the difficulty people often have in understanding documents and were familiar with various strategies for overcoming these difficulties.

Each document was edited individually by each of the designers, and then reviewed in a joint session in which a composite revision was written. After each editor had individually reviewed the composite, the group met again to work out a final joint draft. Over one hundred hours were devoted to the revision of the four documents.[3]

The documents

Four documents were revised in this study: a guarantee, an explanation of Workers' Compensation benefits from an employee handbook, a consent to surgery form, and the first page of a car insurance policy. All the documents were one page long. Their main function was to give information to the user, rather than to gather information from the user. They were documents that the user would have available to refer to later; in fact ordinarily the document might not be read until it was needed. Therefore it was important that the information be presented in a way that could be easily accessed and understood but not necessarily remembered. Three of the four original documents and all four revisions are shown in the Appendix. Fictitious company names have been inserted on the originals.

The original guarantee and the explanation of Workers' Compensation benefits were documents that people familiar with our project had called to our attention as needing improvement. The original consent to surgery form is typical of those being used by Pittsburgh area hospitals. (The revised consent to surgery form was the result of a project we conducted with the assistance of lawyers who represent several area hospitals [Janik, Swaney, Bond, & Hayes, 1981].)

We included an insurance policy because of the legislation that exists on the readability of insurance policies. According to a 1979 survey only three states had passed laws or regulations on the readability of consumer contracts; but twenty-five states had made such laws or regulations on insurance policies and six others were considering such legislation (Pressman 1979). The original car insurance policy is that of a reputable company and is similar to many policies in current use. It contains complex language, extremely long sentences, tiny print, and too many type faces.

The revisions

Many of the changes which our designers recommended may be characterized as applications of standard editing principles. These included:

1. Breaking long sentences into shorter ones (On the car insurance policy, the average sentence length went from 42.5 words to 22.1);
2. Deleting unusual or unnecessary words such as "therein," "accrues," "hereunder," and "deems" or replacing them with common words;
3. Breaking the material into meaningful sections with descriptive headings;
4. Putting long conditionals into list format;
5. Deleting unnecessary information;
6. Replacing vague language with concrete language and examples;
7. Modifying the graphic design by using a size of type which can be easily read and by using white space and indentation to reinforce the organization.

Other changes which the designers recommended, especially changes in organization, are harder to characterize as applications of standard editing principles. Rather, they appear to reflect problem solving stimulated by the peculiar demands of the topic and the audience. For example, information in the hospital consent form was reorganized into four separate sections to highlight the patients' options.

The subjects

Since the users of the documents revised in this study would be people from the general public who, mostly, are out of school and are working, we recruited our subjects from among adult acquaintances and at neighborhood laundromats. Laundromat customers are more representative of the population as a whole than college students are, and they had time to devote to our experiment without being inconvenienced. We felt that the advantage of using appropriate subjects outweighed the disadvantages of the testing situation (noise and lack of privacy).

The subjects were paid $3 for completing the experiment, which took approximately half an hour. In all, 12 men and 12 women, ranging in age from 19 to 69 (median age 31), took part. The median age of the U.S. population is 29.7. Our subjects' education ranged from 12 to 17 years or more, with a median of 14 years. The median schooling of the U.S. population which is 25 years and older is 11.7 years (for the population 14 years and older, the median is 12.4 years). As a group our subjects were very close in age to the population as a whole, but had a little more formal education.

The procedure

Each subject received each of the four documents. The documents were randomly assigned to subjects, but distribution was planned so that each subject received two originals and two revisions. With each document we gave the subjects a page of text which described a situation to which the document was relevant (for example, "You purchased a tool from a door-to-door salesman on August 25 . . ."). In addition, the subjects were presented a list of six written questions. In order to answer these questions the subject had to apply the information in the document to the situation described. We chose the situations and the questions in an effort to approximate actual situations in which such documents might be used.

We also recorded whether subjects had ever had experience with similar documents by asking them four questions at the end of the experiment (for example, "Have you ever collected Workers' Compensation benefits?").

Subjects were permitted to read the materials in any order, and they were permitted to refer to the document as they answered the questions. They were timed from the moment they were given the document until they completed answering the six questions, but they were told that accuracy was more important than speed and that they could take as long as they wished.

Results

Table 1 shows the results of the experiment. For all four documents there was no significant difference in performance among subjects due to age, sex, education, or prior experience with similar documents.

On the average, subjects took less time to read the revisions of documents 1 and 4 than to read the originals. For documents 2 and 3, reading time was about the same for original and revision.

On the first three documents people using the revisions were able to answer more of the questions correctly (although the difference was significant only for documents 1 and 2). On the fourth document, the revision actually had a higher error rate (50% versus 40%), although this difference was not sig-

Table 1. Results of Experiment 1

Document	*Total Errors* Original	Revision	*Significance by Mann-Whitney U Tests*	*Total Time* Original	Revision	*Significance by Mann-Whitney U Tests*
1	33	9	P < .01	51′ 3″	30′ 15″	P < .002
2	32	19	P < .05 by 1 tailed test	85′ 3″	86′ 22″	Not Significant
3	14	8	Not Significant	53′ 59″	54′ 46″	Not Significant
4	29	36	Not Significant	104′ 17″	79′ 2″	P < .10

Note: For each original and each revision the total number of answers was 72 (12 subjects × 6 questions).

nificant. The number of errors made with each document is shown in table 1.

Note in table 1 that only document 1 (a guarantee) showed a significant improvement in both error rate and time on the revision.

The document for which the revision yielded more errors than the original was the car insurance policy. Figure 1 shows the questions we asked our subjects about the policy and the correct answers. At the bottom are the subjects' incorrect answers. On five of the questions the number of errors was nearly equal on the original and revision. Question D accounted for most of the additional errors on the revision. No one who received the original missed question D, but five people with the revision missed it. On the original the statement which provided the answer to the question read, "reasonable expenses . . . not to exceed $50 per day." The revision read, "up to $50 per day for reasonable expenses . . ." (The wording had been changed in order to have parallelism, so that each item in the list would begin with an amount of money.) Everyone who missed the question overstated the amount.

Many of the missed questions involved math. Although the operations required were very simple, many people find it hard to apply information given in prose to solving math problems. In studying the wrong answers and looking at the sheets on which the subjects did the math, we found that the wrong answers were not a result of calculating mistakes. Rather, the subjects often were not able to determine what amount or operation to use.

However, the fact that math was involved did not fully explain why the revision was not effective. Wrong answers indicated that the subjects were not comprehending basic facts which the policy should have made clear.

EXPERIMENT 2: READING PROTOCOLS

At this point we had a document to which all of the standard editing techniques had been applied and on which the readability score had been lowered, yet the revision was not any easier for people to understand. We needed a technique which would pinpoint the reasons for the difficulties. We judged that reading protocols as applied by Flower, Hayes, and Swarts (1980) were most likely to yield the information we needed.

4. You were in a very serious car accident which was covered by your car insurance policy with Vista Insurance. Under your policy you had purchased Limits of Liability of $100,000 for bodily injuries and $50,000 for property damages (these are the maximum amounts that will be paid for any one accident). You read the policy to find out how much of the settlements and your expenses will be paid by Vista. Shown after the following page is the part of your policy that gives this information.

A. One of the people injured in the accident is suing you for $100,000 for bodily injuries. Vista decides to settle out of court and pay the injured party $75,000, but you prefer to fight the lawsuit in court. Can Vista settle out of court without your permission? yes (X) ___ no ___ can't tell ___

B. The first lawsuit was settled out of court for $75,000. Then a second person injured in the same accident sues you for bodily injuries. He wins his case in court. You are held responsible for paying him $90,000 plus his costs of $10,000 for an attorney. How much of the $100,000 will Vista pay on your behalf? $ (35,000) ___

C. Will the fees for the attorney hired by Vista to handle these two cases by paid by Vista? yes (X) ___ no ___ can't tell ___

D. During the above lawsuit, Vista asks you to be present in court each day. You miss 10 days from your job as a salesman, where you would have earned $60 a day in salary plus about $40 a day in commissions. You spend $5 a day for gas to travel to the trial. How much will Vista reimburse you for these expenses? $ (500 ___ or $50 per day)

E. At the time of the accident you were arrested for reckless driving. Vista refused to obtain a bail bond for you, but said that if you were able to obtain one they would pay up to the maximum amount allowed for bail bond under your policy. You were successful in obtaining a bail bond, which cost $300. Later you were cleared of the charges, but your lawyer's fee was $400. How much is Vista obligated to pay to cover your bail bond and your lawyer's fee for this case? $ (350) ___

F. In the situation described in E, did Vista violate their policy with you when they refused to arrange for a bail bond? yes ___ no (X) ___ can't tell ___

Incorrect answers received:

Original	*Revision*
A. can't tell (4); no	A. can't tell (4); no
B. 25,000 (8); 100,000; all; blank (2)	B. 25,000 (6); 90,000; 75,000; 50; all; blank
C. can't tell (2); no (2)	C. can't tell (3); blank
D. no errors	D. 630 (3); 1050; 1550
E. 50 (2); 100; 500 (2); none	E. 700; 550; up to 550; 500; 300 (2); 250; blank
F. yes (2)	F. yes (2); can't tell

Figure 1. Questions asked about the car insurance policy, with correct answers in parentheses.

The subjects

As subjects for the protocols we used three men and nine women who had not taken part in experiment 1. Their ages ranged from 20 to 34 (median 25-1/2), and their education ranged from 12 to 17+ (median 16). Nearly all of them had more formal education than the general population, yet as reported below, they had the same problems with the documents as the subjects in experiment 1.

The procedure

Subjects were given one of the versions (the original or the revision) of the car insurance policy. They were instructed to read the document aloud into a tape recorder, verbalizing their thought processes as they worked. (The experimenter answered no questions and made no comments during the experiment.)

After they had read through the document, the subjects were given the six questions that had been used in experiment 1. They were instructed to answer the questions aloud, explaining where in the policy they were finding the information they needed.

We took three protocols of people reading the original document and two of people reading the first revision.

Subjects reading the original often pointed out problems which the designers had already dealt with in their revision—problems such as extremely long sentences. However, many of the problem areas in the original were still problem areas in the revision. The document design experts were given transcripts of the reading protocols, and they then produced a second revision. We then took four protocols of subjects reading the second revision. The protocols revealed a few areas that still needed improvement. Those areas were modified by our experts to produce a third and final version. Finally, we took three protocols of that revision; these protocols did not identify any new problems. The final revision is shown in the appendix.

In this study we were interested only in determining whether reading protocols could be of value to editors. We did not study the analytical stage in which an expert editor reviews both the reader's statement, "I don't know what that sentence means," and the structure or context of the offending sentence to decide which of many alternative fixes is the appropriate one to make. In future studies we will take editing protocols, as well as reading protocols, so that we can analyze how editors use the reading protocols and how they arrive at solutions for the readers' problems.

The final revision meets the legal requirement for readability of insurance policies sold in Pennsylvania, which is a minimum Flesch Reading Ease score of forty. A readability formula is a mathematical equation that is used to predict the relative difficulty a reader will have with a specific written text. All of the formulas that are in wide use are based on only two criteria—sentence length and word difficulty (measured by the number of syllables per word or by word frequency). Although many experts have pointed out the limitations of readability formulas (Redish, 1979; Huggins, 1977; Shuy & Larkin, 1978; among others), writers of insurance policies must check the readability of their work against the legal requirements in the states where those policies are sold. Of the twenty-five states that have laws or regulations on the readability of insurance policies, ten specify a minimum Flesch Reading Ease score as their criterion (Pressman, 1979).

Findings

We learned from the reading protocols that the results of our first experiment were not misleading—people do have real trouble understanding insurance policies.

In earlier studies (Bond, Hayes, & Flower, 1980; Flower, Hayes, & Swarts, 1980) it was found that when people read a difficult text they exercise a number of strategies for trying to understand that text. Likewise, in the protocols collected for this study, we often saw subjects using or trying to use various comprehension strategies. Sometimes the subjects were successful; sometimes not. When they were successful, they gave us valuable ideas about how to make that area clearer to others. When their strategies failed to lead them to an understanding or even led them to a misunderstanding, they showed us areas that crucially needed to be revised.

Instantiation. In the study mentioned above, Flower et al. found that readers attempt to build scenarios, that is, to organize information around human agents and their actions. They also found that readers attempt, to varying degrees, to instantiate those scenarios, that is, to provide the context and concrete instances that make a general rule more understandable. Readers (and writers) can instantiate a scenario by providing extended examples or cases, by interpolating short examples into the prose, or by using details and specifics to define the meaning of a term. Following are some examples of strategies our subjects used:

a) organizing information around human agents and their actions:

S1: For this part I get out of it, ah, me being the subscriber, Vista is going to cover me and sort of act as my attorney if there should be any type of claim.

S3: I guess what they're saying is they would follow through on any lawsuits to, I guess, to make sure that they are substantial and that they're supported by evidence.

b) providing extended examples or cases:

S1: But even if, let's say, the damage isn't as great as the coverage you have and something comes up later on, like say, ok, for property damage, say you hit somebody's wall and they found out later on that, you know, it cracked a sewer line or an underground telephone cable . . .

S8: That property damages would be like—the person's car or whatever you hit or whatever. That $100,000 would be for hospital bills and stuff, I guess. The $50,000 would be for the other guy's car or whatever you hit. Oh, that would probably be for lost wages, too, in there someplace.

c) interpolating short examples into the prose:

S7: Collision must be when your car hits something. I suppose if something like a tree fell on your car, that would be what that was [accident not involving a collision].

S8: That must mean a person's regular insurance—like Blue Cross, I suppose, hospitalization.

d) using details and specifics to define the meaning of a term:

S3: [Limit of Liability] I assume the Limit of Liability is the amount you're insured for, although it doesn't say that here at all.

S8: [lawyer fees] That would be to decide whose fault it was.

Flower et al. argue that instantiation is an important technique which should be utilized by editors and writers. In another study which concentrated on the effect of examples, it was shown that examples inserted in text increase comprehension, even though as a result the revised document is longer than the original (Pollchik, 1975).

Examples and other types of instantiation not only clarify difficult prose and make the general more concrete, but they may also help to solve another problem present in many complex documents—information density. Subjects both in experiment 1 and in our protocols often commented on the large amount of information to be grasped even though the document (the original as well as our first revision) was only one page long.

Kintsch and Vipond (1978) have said that the difficulty of a text is affected by the propositional density and the number of new concepts per proposition. In a short span of time people can grasp only a certain number of ideas. When more words are used to express an idea, the ratio of ideas to time becomes lower and thus more favorable and results in faster reading and improved recall (Wright, 1968; Frase & Fisher, 1976). Examples give the reader more information-processing time.

In the final revision, the designers provided four extended examples (which were set apart in boxes); six short examples interpolated into the text; and seven instances of details and specifics to define the meaning of a term.

Subjects often commented that they liked having examples and that the examples in the second and third revision were clear. One subject commented:

S9: And I like the fact that they cook up the examples in such a way as to show you in a way that's not always flattering to Vista. In other words they show you exactly what your costs, uh, responsibilities are, through their examples. They could have made the examples much more dishonest, to look like Vista would pay in every contingency.

Context. Readers bring their prior knowledge of the material and their knowledge of schemes to most reading tasks. Certainly most people have some knowledge about car insurance. Prior knowledge can be extremely valuable. Serious investigations of the effect of the reader's knowledge base are just beginning, but there is some data to support the claim that the more one knows about a topic, the easier one can learn new things about it (Kintsch & Vipond, 1978, p. 32).

Lack of relevant knowledge can cause problems for readers, as can prior knowledge that is fuzzy or applied incorrectly. For example, the portion of the insurance policy we used dealt only with liability insurance, but we found that most people didn't understand the differences between various types of car insurances, usually all covered in one policy.

Although subjects understood from the document that liability insurance pays for injuries and damages to the other party only, they often questioned why their own injuries and damages weren't mentioned. This misunderstanding did not affect the error rate in experiment 1 because the hypothetical situation and questions did not deal with the policyholder's injuries and damages, but it would cause problems in real life. (An advantage of doing reading protocols is finding areas of misunderstanding which would not be covered by questions, either because the number of questions must be limited or because the person making up the questions doesn't realize that there could be a misunderstanding on that particular point.)

Subjects often brought up topics such as deductibles (which do not apply to liability insurance), no-fault insurance (mandatory insurance required in Pennsylvania and some other states), and time limits. None of these were mentioned in the section of the document that the subjects were reading.

People often think they have heard or read what they expected to. In Davis's study (1977) mentioned earlier, many people couldn't believe that they would be in default on a consumer credit contract if they damaged the goods, even though they kept up the payments. Readers know from experience that most creditors do not care whether or not goods bought on credit have been damaged as long as they are paid for, so it is difficult for them to realize that they would be in default.

Of course, it is impossible to prevent all of the possible misunderstandings which readers might have. However, by being aware of problems which trouble readers frequently, an editor can help to solve those problems.

Inferencing. There was evidence in the protocols of readers making inferences. Examples of correct inferences are:

> *S2:* It sounds as if that means in fact in doing their own investigation, if it seems more beneficial to them to pay, or more expedient, they'll do that. It's up to them cause they're the ones covering all settlements.
>
> *S7:* I suppose someone reading this policy would know how much they should have, that the salesman would explain that to them.

But what was more revealing were wrong inferences or inferences that subjects failed to make (which showed up when they answered the questions). Examples of wrong inferences are:

> *S2:* It sounds like they'll pay the total amount of the bond, um, but only $50 of the lawyer's fee.
>
> *S8:* "Interest on unpaid settlements—for which Vista"—it must mean money those people aren't using.

Among the inferences which subjects often missed were:

- that the phrase "shall defend any suit" means hiring and paying for lawyers;

- that the statement "and may make such investigation and settlement of any claim or suit as it deems expedient" means that the company can settle out of court without the policyholder's permission (our question A).

Failing to make inferences is in part a reading problem that a document can't always solve. But if the inference is crucial and if there is evidence that many readers are failing to see it, the writer should be more explicit. This was done in the final revision.

Attention management. Readers sometimes remind themselves to summarize or to reread something that's not clear to them. Some examples are:

> *S4:* The Limits of Liability is the only thing [I don't understand]. I should probably go over it one more time.
>
> *S9:* I'm going to go back up here and get my bearings.
>
> *S6:* I'd have to reread that to get it all.
>
> *S9:* Ok, I think I've read enough to understand, um, the form of that.
>
> *S3:* Um, boy, I really have to sift through this paragraph to figure out what's going on. I myself am really reluctant to sift through it at all, it's really (laugh) dull, that part of it. [In this example the subject realized what she needed to do, although she didn't follow through.]

A reading error which often surfaced, and which was crucial to understanding a whole section, was understanding the relationship between Limits of Liability and Supplementary Payments. All of the versions, including the original, stated that payment for the items listed under Supplementary Payments (such as lawyer fees, bail bonds, etc.) would not be deducted from the Limits of Liability (the amount of insurance purchased to cover the other party's bodily injuries and property damages); however, once the Limits of Liability have been paid in full, no more Supplementary Payments will be made to cover the costs involved in lawsuits.

We found that subjects usually read the one-line explanation at the beginning of the Supplementary Payments section quickly without dwelling on it. On the original the line said, "To pay, in addition to the applicable Limits of Liability:". The first revision said, "In addition to the Limits of Liability, Vista agrees to pay:". One subject, who made a mistake but later realized it, summed up her problem:

> *S1:* Hmmm, in rereading this, I see that the lawyer fee wouldn't come out of the Limits of Liability because it says these are in addition to the applicable Limits of Liability. It's tough the first time you read a document to get everything out of it and to think of every question you're going to have about it, unless you've thought about it and read through policies before and already have written down questions.

Others were confused by the fact that although Limits of Liability and Supplementary Payments are two different things, there is a relationship between them. Sometimes, as shown below, they realized their confusion:

> *S9:* Why is Vista going to pay all of the expenses of the lawsuit, just because the Limits have not been exhausted? I mean there's some kind of connection between the Limits of Liability, and, uh, paying the expenses of

the lawsuit. . . but it's not—I wish they would have made that relationship a little more clearer.

As with inferences, the writer can't solve all of the readers' problems. However, where there is evidence of many people making the same mistake, an editor should try to correct the problem. The one-line explanation at the beginning of the section clearly was insufficient. In the final revision the designers repeated the distinction at several points, made the distinction clear in examples in each section, and specifically cautioned people to keep it in mind.

Dealing with lexical problems. The following comments are examples of subjects attempting to deal with lexical problems, that is, problems with words and vocabulary:

S7: I guess I don't know what an immediate family member is exactly. Is it my grandmother?

S7: I don't know what a premium is, but I suspect most people would, I guess.

S7: I don't have any idea what appeal bonds and, uh, other bonds are.

S6: I don't know what they mean by "been exhausted."

The wrong answers we received to the questions we asked pointed up another lexical problem. Quite a few subjects thought that if the policy paid "up to $500" for a certain item, the policyholder would get reimbursed $500 even if the item cost less. The same mistake was made occasionally when the document read "not to exceed $500." Also, in dealing with a paragraph which mentioned two such categories, subjects often did not keep the two amounts separate; in answering the question, they totaled the two amounts actually spent and then checked to see if the total was under the total of the two maximums allowed.

Many of the lexical problems on the original had already been dealt with by the designers on the first revision. In the longer second revision, some new terms were introduced without adequate definition. The final version contained no undefined terms which gave subjects trouble. The "up to $———" problem was solved by explicitly stating what would be paid if the amount went over the maximum and if the amount were under the maximum, and by giving examples of both conditions.

Use of structure. Subjects occasionally commented on structure (the organization, the headings, etc.). For example:

S3: It's a relief to see this smaller paragraph.

S1: The only thing is under part 1 they tell you to "See part VI for definition of words in bold type." I think defining them as they come up is a better bet.

S4: [about the first revision] It emphasizes all the points. It goes pretty well in order. It tells you your limits in the beginning and then all the other factors.

S6: I thought it makes it a little easier to understand in the beginning, how it's broken down into—oh—a little summary of the different points.

The first revision had solved the structural problems of the original, and most comments on structure in the revisions were favorable. The one negative feature subjects pointed out was frequent references to other sections. Some subjects found this distracting. Although reorganization eliminated a few of these references, the designers decided to keep the rest because in real life a policy is rarely read from cover to cover; instead people look up applicable sections. Therefore we felt it was necessary to refer the reader to any other section that gave information related directly to the section being read.

Overall, it is interesting to note that in their first revision, our designers dealt primarily with lexical and structural problems. In general, though, they did not discover the difficulties which the subjects were having with instantiation, context, inferencing, and attention management until they examined the reading protocols.

EXPERIMENT 3: REPEAT OF WRITTEN TEST

Based on reading protocols, we felt that the third revision would be clear to most people. We then repeated the written test on the insurance policy given in experiment 1.

We recruited 12 new subjects (6 women and 6 men) from among acquaintances and at neighborhood laundromats. Their age range was 20 to 56, with a median of 25. Their years of formal education ranged from 12 to 18, with a median of 13-1/2.

As shown in figure 2, the error rate was significantly lower than with the original document used in experiment 1. In experiment 1 the original had a 40% error rate and the revision 50%. In this experiment the revision had an error rate of 22%.

The average time it took subjects to read the document and answer the questions increased, but not in proportion to the increase in the document's length. The final revision contained nearly four times as many words as the

Version	Total Errors	Significance by Mann-Whitney U Tests	
Original Insurance Policy	29	Difference Not Significant (Original / First Revision)	Difference Significant at .05 by 1 tailed test (Original / Final Revision)
First Revision	36	Difference Significant at .01 (First Revision / Final Revision)	
Final Revision	16		

Figure 2. Results of experiment 3.

original, but it only took subjects an average of 2-1/2 times as long to read the revision as it had the original.

Although the increased length may seem alarming at first glance, there are three mitigating facts:

- As mentioned above, the amount of time it took subjects to read the document and look up facts did not increase in proportion to the increase in length;
- Subjects did not complain about reading the final revision, whereas the subjects who received the original in experiment 1 and in the protocols made many negative comments;
- An insurance policy is a reference document which typically is not read cover to cover. Instead people read only those sections which contain the information they need in a particular instance.

It can be argued that, because we gave the protocol subjects the same questions we gave to the subjects in experiments 1 and 3, the increased comprehension shown in experiment 3 was due to our document design experts making editing changes to fit the questions. However, we believe that the entire content is much clearer in the final revision than in the original document and that similar results would have been obtained if different questions had been used in the protocols. Our document design experts tried to resolve all of the problems evidenced by readers, whether or not that particular area was covered by the questions (and many weren't); the fact that readers had no major problems with the final revision indicates that our experts were successful. While the protocols from this study suggest that it is very useful to have reading protocol subjects answer questions, in future studies we will use a different set of questions (unseen by the editors) to test for comprehension.

FUTURE RESEARCH

We believe that reading protocols have already shown themselves to be a very valuable editing tool. However, as an aid in applying protocols in practical situations, it would be useful to have answers to the following questions:

1. How many protocols are necessary to revise a document adequately? Protocols are time-consuming to collect and transcribe, and therefore expensive. However, because protocols can reveal detailed and valuable information about how users deal with a document, it may not be necessary to spend additional time and money on other types of evaluation. (In any event it should be remembered that the expense of protocols and other tests may be minor compared to the cost of errors made hundreds or thousands of times due to a poor document in wide use.)

2. How can questions best be used with protocols? In our experiment we asked the subjects to answer questions as part of the protocol (after they had given a reading protocol of the document itself). The answers were very valuable in bringing to light misunderstandings which the subjects were not aware of as they read.

3. Will all editors benefit from collecting protocols? Certainly protocols can increase an editor's awareness of readers' problems. But although subjects occasionally offer valuable suggestions on how to change the document, largely it is up to the editor to solve the problems the protocols point out. As we mentioned earlier, we need to know how editors move from awareness of a problem to its solution.

4. Can protocols be used in training editors? If so, will the training apply only in editing subject matter similar to that covered in the protocols, or will it generalize to other types of material?

Our future research with both reading and editing protocols will attempt to answer these questions.

CONCLUSION

Our experiment demonstrates two significant things for document designers. The first is the need to test a document on people who are typical of real users. Although editors who are familiar with clear writing guidelines can make a great improvement on many documents, it is very dangerous to rely solely on guidelines, readability formulas, and intuition. It is not necessarily easy for editors, who of course are familiar with their subject matter, to judge what other people may understand or misunderstand; therefore it is crucial that the editor go one step further and test the revision on members of the intended audience. If problems are found and a new revision is drawn up, then that revision should be tested also. The solution which an editor believes will work may not always accomplish its objective.

The second thing shown in our study is the benefit of using reading protocols as an editing tool. The use of reading protocols to improve documents makes good sense. It removes the editor and experimenter from the situation and lets the reader deal directly with the document. Any trouble spots that exist are likely to be caught, rather than just those items an experimenter happens to ask questions about. Our subjects gave us important information about the problems they had with the document and with the subject matter generally. Based on their comments our designers were able to produce a revision which we believe people will be willing to read and which they can understand.

This report was part of the work done under Contract #400-78-0043 of the National Institute of Education. The project officer was Dr. Candace Miyamura of the Teaching and Learning/Reading and Language Group. This report does not, however, necessarily reflect the views of the funding agency.

This paper was presented at the annual meeting of the American Educational Research Association, Session 15.27, "Making Public Documents Understandable: Research on Problems and Solutions," Los Angeles, April 1981.

1. By standard editing procedures we mean the use of editing guidelines, such as those provided by Strunk and White, together with the editor's intuition about what will be clear and acceptable to the reader, to produce a more intelligible text.

2. In the cloze test, words are deleted—usually every 5th word—and the subject is asked to supply the missing words.

3. Later, for comparison we asked two other expert editors (English teachers at the university) to revise the car insurance policy. Their revisions were similar to the first ones our designers produced.

REFERENCES

Bond, S.; Hayes, J.; and Flower, L. "Translating the law into common language: A protocol study," Document Design Project Technical Report No. 8, Pittsburgh, Pa.: Carnegie Mellon University, April 1980.

Charrow, V.; Holland, M.; Peck, D.; and Shelton, L. "Revising a Medicaid recertification form: A case study of the document design process." Washington, D.C.: American Institutes for Research, 1980.

Davis, J. "Protecting consumers from overdisclosure and gobbledygook: An empirical look at the simplification of consumer-credit contracts." *Virginia Law Review,* 1977, *63*(6), 841–907.

Flower, L.; Hayes, J.; and Swarts, H. "Revising functional documents: The scenario principle," Document Design Project Technical Report No. 10. Pittsburgh, Pa.: Carnegie Mellon University, 1980.

Frase, L., and Fisher, D. "Rating technical documents." Case 25952, Memorandum for File. Piscataway, N.J.: Bell Laboratories, 1976.

Huggins, A. W. F. "Syntactic aspects of reading comprehension," Technical Report No. 35. Cambridge, Mass.: Bolt Beranek and Newman, Inc., 1977.

Janik, C.; Swaney, J.; Bond, S.; and Hayes, J. "Informed Consent: Reality or Illusion?," Document Design Project Technical Report, Pittsburgh, Pa.: Carnegie Mellon University, 1981.

Kincaid, C.; Kincaid, J.; and Lewis, R. "Even loan forms can be readable." *Human Factors Society Bulletin,* 1977, *20*(11), 8–9.

Kintsch, W., and Vipond, D. "Reading comprehension and readability in education practice and psychological theory." Paper presented at the Conference on Memory, University of Uppsala, June 1977. In L. Nilsson (Ed.), *Proceedings of the Conference on Memory.* Hillsdale, NJ: Lawrence Erlbaum Associates, 1978.

Pollchik, A. "The use of embedded questions in the facilitation of productive learning." Doctoral dissertation, Vanderbilt University, 1975.

Pressman, R. "Legislative and regulatory progress on the readability of insurance policies." Washington, D.C.: American Institutes for Research, 1979.

Redish, J. C. "Readability." In D. A. MacDonald (Ed.), *Drafting documents in plain language* (Handbook A4-3034). New York: Practising Law Institute, 1979.

Shuy, R. W., and Larkin, D. L. "Linguistic considerations in the simplification/clarification of insurance policy language." *Discourse Processes,* 1978, *1,* 305–21.

Wright, P. "Reading to learn." *Chemistry in Britain,* 1968, *4,* 445–50.

APPENDIX

GUARANTEE

____
Date

If you are not satisfied with the material for any reason, you may return it without any penalty or obligation, within thirty days from the above date.

If you wish to return the material, notify the Seller by telephone or mail a signed and dated copy of this guarantee or any other written notice, or send a telegram, not later than midnight of ________________. You must make the material available to the Seller at your residence; or you may if you wish return the goods at the Seller's expense and risk.

If you return the material, you may choose to receive a refund or, at the Seller's discretion, a replacement. If you choose a refund, any property traded in, any payments made by you under the contract or sale, and any negotiable instrument executed by you will be returned within 10 business days following receipt by the Seller of the returned material, and any security arising out of the transaction will be cancelled.

If you do not agree to return the goods to the Seller or if the Seller does not pick them up within 20 days of notification, you may retain or dispose of the goods without any further obligation.

I wish to return the material purchased under this transaction.

____
Date

____
Buyer's Signature

I hereby acknowledge receipt of two (2) completely filled in copies of the above guarantee.

____
Customer's Signature

____
Date

Original of document 1. *Readability:* Flesch—35.201 difficult (academic journals); Recalculated Flesch formula—Grade 16.7; Fog index—Grade 18.5. *Average sentence length:* 33.5 words.

GUARANTEE

Terms of the Guarantee

1. We guarantee the material listed below for a period of thirty days from the date of sale.
2. If the material breaks or if you are not satisfied with it for any reason, you may return it to us for a full refund or replacement.
    a. If you desire a replacement, we will try to provide it if possible. However, if a replacement is not available, we reserve the right to refund your money.
    b. If you desire a refund, we will always provide it.
3. If the material is broken due to misuse or vandalism, this guarantee does not apply.

How to Return the Material

You may return the defective material in either of the following ways:

1) Send the material parcel post collect, postmarked no later than 30 days from the date of sale. Include a note asking for either a refund or a replacement.
2) If you prefer that we pick up the material, notify us by telephone within 30 days from the date of sale to arrange a time when we may pick up the material.

Material ______________________________

Date of Sale ______________________________

______________________ ______________________

Customer's Signature Salesman's Signature

Revision of document 1. *Readability:* Flesch—60.53 fairly difficult (quality magazines); Recalculated Flesch formula—Grade 9.0; Fog index—Grade 12.0. *Average sentence length:* 16.9 words.

WORKERS' COMPENSATION

The Pennsylvania Workers' Compensation Law provides compensatory payments to employees for disabilities that are related to employment. The law defines benefits which are intended to cover most of the employee's resulting economic loss, to provide income benefits to survivors, and to cover medical expenses. The law also requires Pennsylvania employers to have insurance (or be self-insured) for these liabilities. Therefore, Fitch Manufacturing provides benefits for injuries or disease resulting from work at Fitch Manufacturing under a workers' compensation and employer's liability policy with the Pennsylvania Manufacturers Association Insurance Company.

Workers' Compensation benefits are of two kinds: 1) cash benefits including income benefits for temporary or permanent disabilities, impairment benefits for partial (scheduled) disabilities, and 2) income benefits for survivors in cases of fatalities. There is a seven day waiting period for income benefits. However, if the disability lasts longer than fourteen days, income benefits are then paid retroactively to the date of the injury. The income benefit in Pennsylvania is set at 66-2/3 percent of the disabled employee's wages. The law also designates minimum and maximum weekly payments based on the claimant statewide average weekly wage (the current minimum is $66 per full week and the maximum is $242 per full week). Medical benefits are not limited in any way by Pennsylvania law for a work related injury or occupational disease; medical benefits begin immediately.

The Pennsylvania law also establishes a time limit within which injuries or occupational diseases must be reported to an employer. Under this provision, compensation will be paid based on the date of the injury only if the injury is reported to Fitch Manufacturing within 21 days of its occurrence. After 21 days, compensation will begin based on when the report is made. The law bars compensation for any injury or disease reported more than 120 days after the occurrence, and it bars all claims in three years after the injury, death, or last compensation payment. It is important, therefore, to be sure that the reporting requirements described in the section on safety policy are observed in the event of any work-related injury or illness.

Fitch Manufacturing bears the full cost of this benefit.

Original of document 2. *Readability:* Flesch—20.928 very difficult (scientific journals); Recalculated Flesch formula—Grade 15.2; Fog index—Grade 17.1. *Average sentence length:* 19.9 words.

COMPENSATION FOR JOB-RELATED DISABILITIES

Types of Payments

If you are disabled by injury or illness caused by your job, Fitch Manufacturing will pay Workers' Compensation benefits as follows:

- All related medical expenses
- Income as follows:

  If you are off work one week or less, you will receive no payment;

  If you are off work more than one week but less than three weeks, you will be paid a portion of your wage for the total number of days missed minus the first five days;

  If you are off work three or more weeks, you will be paid a portion of your wage for all of the days of work missed.

  The portion which you will receive is 2/3 of your gross wage, except that you can receive no less than $66 per week and no more than $242 per week for each full week you qualify for compensation.*

- A monthly benefit or a lump sum settlement if the injury or illness results in partial or total permanent disability (even if you are able to return to work). Fitch Manufacturing will decide which method of payment (monthly benefit or lump sum) you will receive.
- Payments to your survivors if the injury or illness results in your death.

Deadline for Filing

In order to receive full compensation, a claim must be filed with your immediate supervisor within 21 days of the injury or illness. If a claim is filed between the 22nd and the 120th day, compensation will be based on the date the report was filed instead of the date of the injury or illness. Compensation will not be paid if a claim is filed more than 120 days after the injury or illness.

*For example, if your gross wage (wage before taxes) is $300 per week, 2/3 of your wage is $200. This amount is between $66 and $242, so this is the amount you would be paid for each full week of work you are eligible for benefits. If you are eligible for only 2 days, you would receive $80 ($200 per week ÷ 5 days = $40 per day; $40 × 2 days = $80).

Revision of document 2. *Readability:* Flesch—59.21 fairly difficult (quality magazines); Recalculated Flesch formula—Grade 10.4; Fog index—Grade 12.8. *Average sentence length:* 21.6 words.

PATIENT: ______________________

DATE: ______________ TIME: ______________ A.M. P.M.

CONSENT TO OPERATION OR OTHER MEDICAL PROCEDURE

I hereby acknowledge and understand that the nature of the procedures authorized by this form, the possible alternative methods of treatment, and risks involved or the probabilities of success or failure have been fully explained and disclosed to me personally by Dr. ______________.

1. I hereby authorize the performance upon ______________________
(Myself or Name of Patient)

of the following procedure ______________________
(Name or Description of Operation or Other Procedure to be Performed in the Language of Laymen)

2. I hereby authorize Dr. ______________ and/or such assistants, including resident physicians, as may be advisable or necessary, to remedy the condition or conditions which are indicated by the diagnostic studies already performed.

3. I have also been informed that there are other risks such as severe loss of blood, infection, cardiac arrest, etc. that are attendant to the performance of any surgical procedure. I acknowledge that no guarantees have been made to me concerning the results of the operation or procedure.

4. I, Dr. ______________, affirm that I have personally communicated to the patient the information referred to above, have offered explanations to the best of my ability, have answered the questions raised by the patient and have explained the contents of this consent form to the patient who has signed it in my presence.

______________ ______________________

(Date) (Signature of Physician)

5. I, ______________________, (Patient or Person Authorized to Consent for the Patient) attest to the fact that I have been given the opportunity for questions and answers regarding the condition(s) and procedure(s) outlined in this consent form and the contents thereof have been fully explained to me. The blanks appearing on this form have been completed or deleted prior to my signature.

____

(Signature of Patient or Person Authorized to Consent for Patient)

Witness

____

Original of document 3. *Readability:* Flesch—29.73 very difficult (scientific journals); Recalculated Flesch formula—Grade 17.0; Fog index—Grade 20. 4. *Average sentence length:* 31.8 words.

<table>
<tr><td></td><td>City Hospital
Consent to Surgery</td></tr>
<tr><td>Part 1
Planned
Treatment
To be completed by the doctor, in common terms, before the patient signs this form.</td><td>1. Dr. ____________ has explained to me that:
a. I have the following condition: ____________
____
____
b. my condition needs to be treated because: ____________
____
____
____
2. I authorize Dr. ____________

The operation to be performed on me is as follows: ____________
____
____
3. The doctor has explained that:
a. the following risks are involved in the operation: ____________
____
____
____
b. all surgery involves risks such as severe loss of blood, infection, and heart stoppage;
c. the surgery will be done in a responsible manner by a licensed physician. No guarantee has been made that the surgery will improve my condition;
d. my condition could be treated in the following other ways which are not being used on me at this time:</td></tr>
<tr><td>Part 2
Additional
Surgery
To be completed by the patient.</td><td>My doctor has explained to me that sometimes during surgery it is discovered that additional or other surgery is needed. If I need additional or other surgery during my operation,
☐ I authorize the doctor to proceed.
☐ I do not authorize the doctor to proceed.</td></tr>
</table>

Revision of document 3. *Readability:* Flesch—49.22 difficult (academic journals); Recalculated Flesch formula—Grade 9.8; Fog index—Grade 13.5. *Average sentence length:* 14.1 words.

<table>
<tr>
<td>Part 3
Disposition of Tissues
To be completed by the patient.</td>
<td>I request that anything removed from me:
☐ be disposed of by the hospital as usual, or
☐ other: ____________________
____________________</td>
</tr>
<tr>
<td>Part 4
Signatures
To be completed only after Parts 1, 2, and 3 are filled in.</td>
<td>• I declare that I have personally explained the nature of the patient's condition, the need for treatment, the operation to be performed, and the risks and alternatives listed in Parts 1, 2, and 3:
☐ to the patient.
☐ to the patient's closest relative or guardian:
Name ____________________
Address ____________________
____________ Phone ____________

I have given the patient or the person named above an opportunity to ask questions which I have answered as fully as possible.
Signature of Doctor ____________________
Signature of Witness ____________________
Date ____________ Time ____________ am pm
• The doctor has explained all the information in Parts 1, 2, and 3, to me and has answered my questions.
I understand that a copy of this completed form will be given to me after I sign it.
Signature of Patient ____________________
Signature of Witness ____________________
Date ____________ Time ____________ am pm
• The patient is unable to consent for the following reason: ____________________
____
____

I therefore give consent on the patient's behalf.
Signature of Closest Relative or Legal Guardian ____________________
Relationship to Patient ____________________
Signature of Witness ____________________
Date ____________ Time ____________ am pm</td>
</tr>
</table>

VISTA INSURANCE CO., INC.
AUTOMOBILE POLICY

Limits of Liability

Vista agrees to pay for bodily injuries or property damages caused by an automobile accident for which you are held legally responsible and which involves a vehicle or driver insured under this policy. The section "Vehicles and Drivers Covered" explains which vehicles and drivers are covered under this policy. The maximum amounts that Vista will pay for injuries or damages occurring in one accident are the amounts of the Limits of Liability purchased under this policy ($100,000 for bodily injuries and $50,000 for property damages).

Lawsuits

Vista agrees to defend, at its cost, any lawsuit for bodily injuries or property damages which would be payable under this policy. Vista may make investigations and settle any claim or suit as it wishes. However, once Vista has paid out in judgments or settlements the amounts of the Limits of Liability, Vista is not obligated to defend any additional suit or to pay any further claims or judgments for bodily injuries or property damages.

Supplementary Payments

In addition to the Limits of Liability, Vista agrees to pay:

—all expenses incurred by Vista in connection with lawsuits that Vista agrees to defend;

—all court costs and attorney fees incurred by you in connection with lawsuits that Vista agrees to defend;

—up to $50 per day for reasonable expenses, including loss of wages or salary, incurred by you due to attendance at hearings or trials at Vista's request (income other than wages or salary, such as tips or commissions, will not be paid);

—up to $500 for a bail bond resulting from an accident or traffic law violation involving an automobile insured under this policy (however, while Vista agrees to pay for such bail bonds, Vista is not obligated to apply for or to furnish them);

—up to $50 in attorney's fees if you are arrested in connection with an accident involving an automobile insured under this policy.

First revision of document 4. *Readability:* Flesch—38.630 difficult (academic journals); Recalculated Flesch formula—Grade 13.4; Fog index—Grade 17.2. *Average sentence length:* 22.1 words.

This policy provides you with several types of car insurance. A brief summary of what is covered by each type of insurance is listed below. Each type is explained more fully in the rest of the policy.

Throughout this policy, "you" refers to the policyholder and all relatives who live in the same household. "Vista" refers to Vista Insurance Co., Inc. "Other party" refers to anyone else.

This policy takes effect ______________________, 19____ and will remain in effect until cancelled by the policyholder or Vista.

Summary of This Policy

Payments Made To	*Type of Accident*	*Name of Insurance*	*Explained In*	*Pays For*
Other Party, You (only for expenses of lawsuits or arrest)	Any accident which results in your being sued	Liability	Part I	—bodily injury, lost wages, and death to the other party; —damage to property belonging to the other party; —certain expenses due to lawsuits; —certain expenses due to your arrest.
You and drivers and occupants of your car	Accident in which all the cars are insured	Personal Injury Protection (often called "no-fault")	Part II	—bodily injury, lost wages, and death to you; —bodily injury and lost wages to anyone else driving or occupying your car.
You and drivers and occupants of your car	Accident in which the other car is uninsured, underinsured, or hit-and-run	Uninsured Motorists	Part III	—bodily injury, lost wages, and death to you; —bodily injury, lost wages, and death to anyone else driving or occupying your car; —damage to your property.
You	Any accident which results in damage to your car, whether or not another car is involved	Automobile Physical Damage, including: collision— comprehensive—	Part IV	—damages to your car and certain other expenses resulting from an impact with another car, an object, or a person; —damages to your car and certain other expenses resulting from an accident which does not involve a collision (for example, if a tree falls on your car).

Family car insurance policy. Only this portion of the policy was used in the experiment. *Readability:* Flesch—44.755 difficult (academic journals); Recalculated Flesch formula—Grade 11.2; Fog index—Grade 12.3. *Average sentence length:* 17.4 words.

Part I—LIABILITY INSURANCE

1. *What does Liability Insurance cover?*

   Liability Insurance covers:

   A) Limits of Liability—payments for bodily injuries and property damages suffered by the *other party* as a result of a car accident covered by this policy (explained in #2). The amounts of these payments are listed in #3.

   The term "bodily injuries" means disabilities suffered by the other party. Payment will be made to cover the cost of the following items, up to the amount of the Limit of Liability:

   —injury;
   —sickness;
   —disease;
   —death;
   —loss of income due to any of the above.

   The term "property damages" means harm to anything belonging to the other party (a car, motorcycle, house, tools, or any other object). Payment will be made to cover the following costs, up to the amount of the Limit of Liability:

   —partial or total destruction;
   —loss of use of the property (for example, cost of a rental car if the other party's car cannot be driven).

   B) Supplementary Payments—payments for certain expenses connected with lawsuits brought against you, as well as certain expenses connected with your arrest due to an accident or traffic violation. The amounts of these payments are listed in #6 A, B, and C.

2. *What accidents are covered?*

   Liability Insurance covers any accident that meets both of the following conditions:

   A) the accident involves either a driver or a car insured by this policy (explained in #7);

   and

   B) you are sued and the court decides that the accident was your fault. (Of course, if the court decides that you were not at fault, neither you nor Vista would have to pay for injuries or property damages of the other party.)

   The Commonwealth of Pennsylvania requires all car owners to carry no-fault insurance (explained in Part II). Therefore, if you are involved in an accident with another Pennsylvania car owner, it is possible that the accident will not result in a lawsuit. In that case, your costs for bodily injuries and property damages will be covered by your insurance (under Parts II and IV). The other party's costs will be covered by his or her insurance.

3. *How much will be paid under Limits of Liability for injuries and damages?*

The maximum amounts that Vista will pay for bodily injuries and property damages *occurring in one accident* are called the Limits of Liability. These are the amounts of coverage that you purchase with your premium (the amount of money you pay Vista each year). You have purchased Limits of Liability of $ 100,000 for bodily injuries and $ 50,000 for property damages.

If more than one person is hurt in the same accident, Vista will pay only a combined total of $ 100,000 for bodily injuries. If more than one car or other property is damaged in the same accident, Vista will pay only a combined total of $ 50,000 for property damages. Of course, one accident may result in both bodily injuries and property damages, and Vista will pay up to the maximum for each.

Example

Suppose that you were involved in an accident which the court decides was your fault. You had purchased Limits of Liability of $100,000 for bodily injuries and $50,000 for property damages. Two people in the other car were hurt. The court awarded the driver of the other car $60,000 for bodily injuries and $6,000 for damages to his car and its contents. His passenger was awarded $50,000 for bodily injuries. Vista will pay to the other party:

Bodily injuries:	
Driver of other car	$60,000
Passenger in other car	$40,000 (you would have to pay the other $10,000)
Property damages:	
Other car	$6,000

You may purchase higher Limits of Liability at any time. But, of course, if you are in an accident, the Limits of Liability which will apply are those you had purchased prior to the accident.

4. *Is there a time limit on payments?*

If an accident occurs during the effective dates of this policy, it will be covered during the entire period that a lawsuit may be legally brought against you, whether or not this policy has been cancelled in the meantime.

5. *What rights does Vista have?*

Vista may make investigations and settle any claim or lawsuit as it wishes. This means that for any accident covered by this policy Vista may gather information about the accident, and it may settle in or out of court for any amount it wishes without your permission (up to the remaining amount of your Limits of Liability). Vista will select the lawyer and pay the lawyer's fees.

6. *How much will be paid under Supplementary Payments for expenses?*

   Vista agrees to pay expenses due to lawsuits, your expenses to attend hearings or trials, and expenses due to your arrest. The amounts that will be paid for each type of expense are explained below. Payments for these expenses are called Supplementary Payments because they are *in addition to* the Limits of Liability. Remember that Limits of Liability and Supplementary Payments are two different things. However, once the Limits of Liability have been paid in full, Vista will not make any more Supplementary Payments to cover lawsuits.

   A. *Expenses due to lawsuits*

      As long as your Limits of Liability for bodily injuries and property damages have not been used up, Vista will make Supplementary Payments to pay in full all expenses connected with lawsuits resulting from an accident covered by this policy. The expenses which will be paid are:

      —lawyer fees;
      —court costs;
      —interest on money which Vista has promised to pay the other party, but has not paid yet;
      —premiums on bonds which must be bought if Vista decides to appeal a court decision;
      —premiums on any other bonds required due to a lawsuit.

      Vista will pay all of the above expenses incurred by:

      —Vista;
      —you;
      —the other party, if the court finds you at fault and orders you to pay the other party's expenses for the lawsuit.

It will sometimes happen that although the Limits of Liability (bodily injuries and property damages) have not been reached prior to the start of a lawsuit, the settlement of that lawsuit will go over the remaining balance. In that case Vista will make Supplementary Payments to cover *all* of the expenses of that lawsuit (lawyer fees, etc.), but it will pay for bodily injuries and property damages only the *amount remaining* in the Limits of Liability. After that, because the Limits of Liability have been used up, Vista will not make any more payments under either Limits of Liability or Supplementary Payments. (However, a new accident would be covered under both types of payments.)

Example

Suppose that you were in an accident which the court decides was your fault. You had purchased Limits of Liability of $100,000 for bodily injuries and $50,000 for property damages. Three people who were in the other car sue you in separate lawsuits and are awarded the amounts shown below.

LIMITS OF LIABILITY:

	BODILY INJURIES			PROPERTY DAMAGES		
	Total	Paid by Vista	Paid by you	Total	Paid by Vista	Paid by You
1st party	60,000	60,000	0	6,000	6,000	0
2nd party	50,000	40,000	10,000	0		
3rd party	20,000	0	20,000	0		

SUPPLEMENTARY PAYMENTS:

	EXPENSES OF LAWSUIT		
	Total	Paid by Vista	Paid by you
1st party	2,000	2,000	0
2nd party	1,000	1,000	0
3rd party	1,000	0	1,000

B. *Expenses to attend hearings or trials*

If Vista requests that you attend hearings or trials, Vista will pay you for reasonable expenses, not to exceed $50 per day. This includes the loss of wages or salary. It does not include the loss of other income such as tips or commissions.

Example

Suppose that Vista requested you to be present at a trial for two days. You ordinarily earn $60 a day, but you were not paid while you attended the trial. Therefore you lost $120 in earnings. You spend $3 on gas to travel to the trial and $3 on lunch each day, an additional expense of $12 ($6 × 2 days). Your total expenses would be $120 + $12 = $132. Vista would pay you a maximum of $50 a day × 2 days = $100. The remaining $32 would come out of your pocket.

If your expenses had been less than $50 a day, Vista would pay the actual amount. For example, if you lost wages of $40 a day and spent $6 a day on gas and meals, your total expenses would be $92 ($46 × 2 days). Vista would pay the full $92.

C. *Expenses due to your arrest*

If you are arrested for a traffic violation which involves a car insured under this policy, Vista will pay the following expenses:

—*In connection with an accident.* If your arrest is in connection with an accident, Vista will pay for:

1) bail bond (your actual cost, as long as it doesn't exceed $500) *and*
2) lawyer fees (your actual cost, as long as it doesn't exceed $50).

If your bail bond costs more than $500 or your lawyer fees more than $50, you must pay the balance. If it costs less, Vista will pay only the actual cost.

Each of these items is considered separately. Therefore, if your bail bond is less than $500, you cannot apply any of the balance toward the lawyer fees (or vice versa).

Do not confuse these lawyer fees with those discussed in Paragraph A. These lawyer fees are for handling any charges the *police* bring against you as a result of an accident. The lawyer fees in Paragraph A are to handle lawsuits brought against you by the *other party* for bodily injuries and property damages.

—*Not in connection with an accident.* If your arrest is not in connection with an accident, Vista will pay for a bail bond only (your actual cost, as long as it doesn't exceed $500). An example of this situation is arrest for drunken driving, even though no accident has occurred.

Although Vista will pay for bail bonds, it is not obligated to apply for or furnish them. You must make these arrangements yourself.

Following is an example which shows all of the types of payments under Liability Insurance.

Example

Suppose that you were involved in an accident which the court decides was your fault. You had purchased Limits of Liability of $100,000 for bodily injuries and $50,000 for property damages.

Three people in the other car were hurt. The driver of the other car was awarded $60,000 for bodily injuries, $6,000 for damages to his car and its contents, and $5,000 for his lawyer fees. The fees for the lawyer hired by Vista and other expenses incurred by Vista were $2,000. Your expenses to attend the trial (lost wages, etc.) for 10 days were $40 a day. At the time of the accident you had been arrested for reckless driving. Your bail bond cost $200 and your lawyer fee cost $175. Vista would pay:

Limits of Liability:	
bodily injuries	$60,000
property damages	$6,000
Supplementary Payments:	
expenses of the lawsuit	$7,000 (5,000 + 2,000)
expenses to attend trial	$400 (40 × 10)
expenses due to arrest	$250 (200 + 50; you would have to pay $125)

Then one of the passengers in the other car sued you and was awarded $50,000 for bodily injuries and $3,000 for his lawyer fees. The fee for the lawyer hired by Vista and other expenses incurred by Vista were $1,000. Your expenses to attend the trial for 5 days were $60 a day this time. Vista would pay:

Limits of Liability:	
bodily injuries	$40,000 (you would have to pay $10,000)
Supplementary Payments:	
expenses of the lawsuit	$4,000 (3,000 + 1,000)
expenses to attend trial	$250 (50 max. × 5; you would lose $10 a day)

At that point, the Limit of Liability for bodily injuries would be used up. The Limit of Liability for property damages would have $44,000 left ($50,000 − 6,000). If the third person in the other car sued you, Vista would not make any more payments for either bodily injuries or the expenses of the lawsuit.

Damages to your own car and injuries to you and occupants ofyour car would be paid by Vista, under Parts II and IV of this policy.

4

CASE STUDIES

Plain English on the Plant Floor

Barry Jereb
The Allen-Bradley Company, Cleveland

Product documentation such as user's guides and programming manuals are key components of the products they support. In 1982 the Allen-Bradley Company, Industrial Computer Group, recognized the need to redefine the way we prepared our manuals – for several reasons: to reduce complaints, to make our products safer and easier to use, and to increase sales. This case study shows how we identified problems with our manual writing, scheduling, and graphics. It shows how we solved these problems by using plain English and clear document design, and identifies the benefits we've gained from our new approach.

At Allen-Bradley Company, Systems Division, we make computers for industrial applications. One way we support these computers is by writing manuals that accompany our products to customers' plants. For many industrial computer companies, writing manuals appears to be an afterthought. In the past two years, however, manuals have become a key component of the products at Allen-Bradley. This paper will show why we decided to improve our industrial computer documentation and what we did to improve it.

Background

At Allen-Bradley the technical writing department was producing a variety of documentation to support our industrial computer systems. This documentation was used by audiences ranging from computer novices such as machinists and factory electricians to computer experts such as computer programmers and software engineers. These documents included installation manuals, maintenance manuals, programming manuals, user's guides, and technical data sheets.

We were unaware of the impact these documents were having on our customers until the Chilton Company completed a study for us in 1982. We had commissioned this study to learn more about the marketplace for our products. To our surprise, the study indicated that product documentation such as user's guides and reference manuals was the *second* most important factor in influencing customers to buy an industrial computer. It was outranked only by quality of product. Other factors, in order, included after-sales service, on-time delivery, knowledgeable salespeople, application assistance, and price.

The results of the study clearly indicated that improved documentation would give us a competitive edge. This discovery led naturally to our taking a closer look at our documentation to see if there were any other indications that our documents needed a change. We found several:

Reprinted by permission of the publisher, *Visible Language*, Rhode Island School of Design, Providence, RI 02903.

Figure 1. Allen-Bradley Company's old layout for manuals.

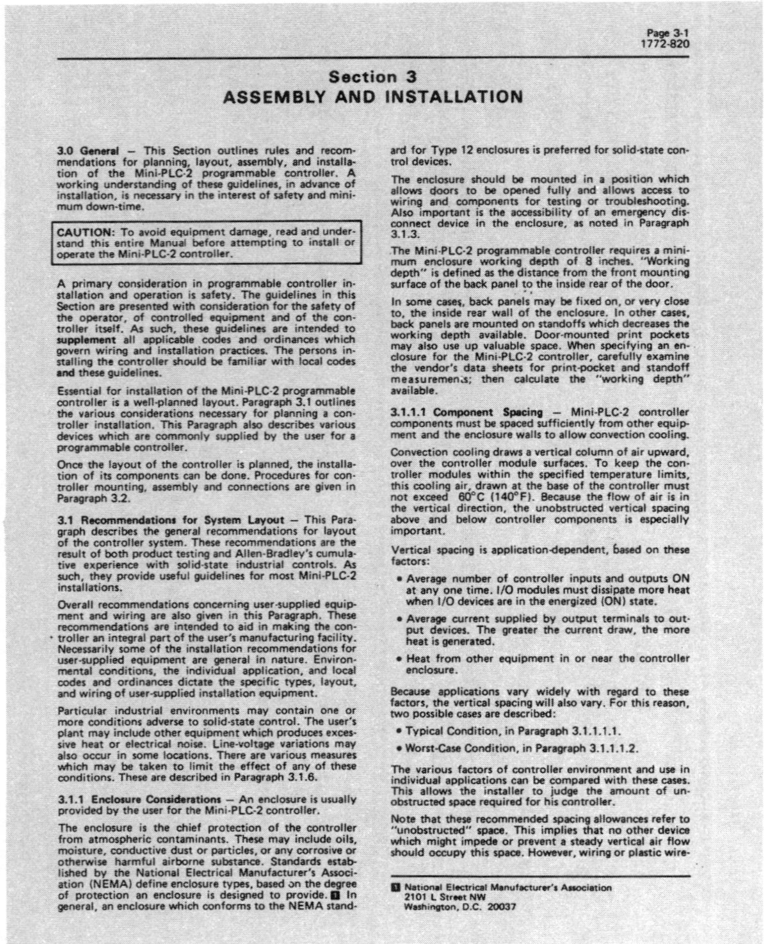

Page 3-1
1772-820

Section 3
ASSEMBLY AND INSTALLATION

3.0 General — This Section outlines rules and recommendations for planning, layout, assembly, and installation of the Mini-PLC-2 programmable controller. A working understanding of these guidelines, in advance of installation, is necessary in the interest of safety and minimum down-time.

CAUTION: To avoid equipment damage, read and understand this entire Manual before attempting to install or operate the Mini-PLC-2 controller.

A primary consideration in programmable controller installation and operation is safety. The guidelines in this Section are presented with consideration for the safety of the operator, of controlled equipment and of the controller itself. As such, these guidelines are intended to **supplement** all applicable codes and ordinances which govern wiring and installation practices. The persons installing the controller should be familiar with local codes and these guidelines.

Essential for installation of the Mini-PLC-2 programmable controller is a well-planned layout. Paragraph 3.1 outlines the various considerations necessary for planning a controller installation. This Paragraph also describes various devices which are commonly supplied by the user for a programmable controller.

Once the layout of the controller is planned, the installation of its components can be done. Procedures for controller mounting, assembly and connections are given in Paragraph 3.2.

3.1 Recommendations for System Layout — This Paragraph describes the general recommendations for layout of the controller system. These recommendations are the result of both product testing and Allen-Bradley's cumulative experience with solid-state industrial controls. As such, they provide useful guidelines for most Mini-PLC-2 installations.

Overall recommendations concerning user-supplied equipment and wiring are also given in this Paragraph. These recommendations are intended to aid in making the controller an integral part of the user's manufacturing facility. Necessarily some of the installation recommendations for user-supplied equipment are general in nature. Environmental conditions, the individual application, and local codes and ordinances dictate the specific types, layout, and wiring of user-supplied installation equipment.

Particular industrial environments may contain one or more conditions adverse to solid-state control. The user's plant may include other equipment which produces excessive heat or electrical noise. Line-voltage variations may also occur in some locations. There are various measures which may be taken to limit the effect of any of these conditions. These are described in Paragraph 3.1.6.

3.1.1 Enclosure Considerations — An enclosure is usually provided by the user for the Mini-PLC-2 controller.

The enclosure is the chief protection of the controller from atmospheric contaminants. These may include oils, moisture, conductive dust or particles, or any corrosive or otherwise harmful airborne substance. Standards established by the National Electrical Manufacturer's Association (NEMA) define enclosure types, based on the degree of protection an enclosure is designed to provide. [1] In general, an enclosure which conforms to the NEMA standard for Type 12 enclosures is preferred for solid-state control devices.

The enclosure should be mounted in a position which allows doors to be opened fully and allows access to wiring and components for testing or troubleshooting. Also important is the accessibility of an emergency disconnect device in the enclosure, as noted in Paragraph 3.1.3.

The Mini-PLC-2 programmable controller requires a minimum enclosure working depth of 8 inches. "Working depth" is defined as the distance from the front mounting surface of the back panel to the inside rear of the door.

In some cases, back panels may be fixed on, or very close to, the inside rear wall of the enclosure. In other cases, back panels are mounted on standoffs which decreases the working depth available. Door-mounted print pockets may also use up valuable space. When specifying an enclosure for the Mini-PLC-2 controller, carefully examine the vendor's data sheets for print-pocket and standoff measurements; then calculate the "working depth" available.

3.1.1.1 Component Spacing — Mini-PLC-2 controller components must be spaced sufficiently from other equipment and the enclosure walls to allow convection cooling.

Convection cooling draws a vertical column of air upward, over the controller module surfaces. To keep the controller modules within the specified temperature limits, this cooling air, drawn at the base of the controller must not exceed 60°C (140°F). Because the flow of air is in the vertical direction, the unobstructed vertical spacing above and below controller components is especially important.

Vertical spacing is application-dependent, based on these factors:

- Average number of controller inputs and outputs ON at any one time. I/O modules must dissipate more heat when I/O devices are in the energized (ON) state.
- Average current supplied by output terminals to output devices. The greater the current draw, the more heat is generated.
- Heat from other equipment in or near the controller enclosure.

Because applications vary widely with regard to these factors, the vertical spacing will also vary. For this reason, two possible cases are described:

- Typical Condition, in Paragraph 3.1.1.1.1.
- Worst-Case Condition, in Paragraph 3.1.1.1.2.

The various factors of controller environment and use in individual applications can be compared with these cases. This allows the installer to judge the amount of unobstructed space required for his controller.

Note that these recommended spacing allowances refer to "unobstructed" space. This implies that no other device which might impede or prevent a steady vertical air flow should occupy this space. However, wiring or plastic wire-

[1] National Electrical Manufacturer's Association
2101 L Street NW
Washington, D.C. 20037

Figure 2. Allen-Bradley Company's new layout for manuals.

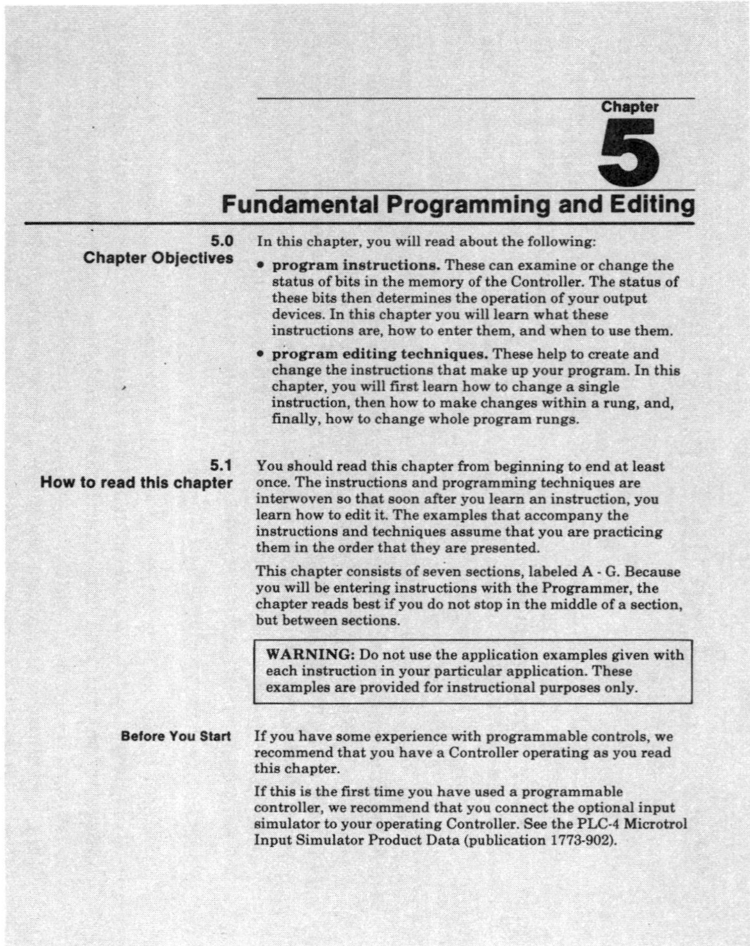

Chapter
5
Fundamental Programming and Editing

5.0 Chapter Objectives

In this chapter, you will read about the following:

- **program instructions.** These can examine or change the status of bits in the memory of the Controller. The status of these bits then determines the operation of your output devices. In this chapter you will learn what these instructions are, how to enter them, and when to use them.
- **program editing techniques.** These help to create and change the instructions that make up your program. In this chapter, you will first learn how to change a single instruction, then how to make changes within a rung, and, finally, how to change whole program rungs.

5.1 How to read this chapter

You should read this chapter from beginning to end at least once. The instructions and programming techniques are interwoven so that soon after you learn an instruction, you learn how to edit it. The examples that accompany the instructions and techniques assume that you are practicing them in the order that they are presented.

This chapter consists of seven sections, labeled A - G. Because you will be entering instructions with the Programmer, the chapter reads best if you do not stop in the middle of a section, but between sections.

WARNING: Do not use the application examples given with each instruction in your particular application. These examples are provided for instructional purposes only.

Before You Start

If you have some experience with programmable controls, we recommend that you have a Controller operating as you read this chapter.

If this is the first time you have used a programmable controller, we recommend that you connect the optional input simulator to your operating Controller. See the PLC-4 Microtrol Input Simulator Product Data (publication 1773-902).

Customer complaints. We had received letters from some customers expressing their dissatisfaction with our documentation.

Legal problems. Because our computers control heavy equipment, an unclear manual could lead to misapplication of a product and could result in equipment damage, injuries to workers, and lawsuits. In addition, an unclear manual could keep customers from getting the system running and thus result in expensive production downtime.

Phone calls. We were forced to create a toll-free telephone center to answer questions about products, many of which should have been answered by our documentation. Staffing a phone center eight hours a day requires time, talent, and money.

Feedback from field. Our salesmen and distributors began to demand better manuals. They wanted to learn about new products so that they could sell them faster. Having good documentation is especially important for international companies like Allen-Bradley; sometimes it is the only information source our international sales groups have.

On the basis of these factors we concluded that our own employees — and even worse, our customers — were dissatisfied with the quality of some of our documentation. Moreover, the Chilton study showed us that our documentation had been hurting us in the marketplace.

Defining areas for improvement

To attack documentation drawbacks, we identified three general areas of our documentation that we had to improve: graphic design, document scheduling, and writing style.

Graphic design

The most evident problems in our documents were graphic. Largely because of a lack of communication between the company's technical writing groups, Allen-Bradley was using three different layouts for technical documentation. As a result, we weren't presenting a consistent corporate identity to our customers. And our page layouts were uninviting and difficult to read (Figure 1). Although this page is similar to layouts used throughout the industrial control industry, we believed that we could do much better.

In developing a new layout we had to remember two things: the constraints on our graphic arts department and our printing budget. We knew the new layout could not be artwork intensive; it had to be easy for our designers to learn and execute. Our printing budget limited our use of color and special paper.

Once we had developed several alternative formats, we took copies to the Customer Training Center and asked our customers to complete questionnaires about the format they preferred. We evaluated the comments and picked a format we thought would be a good starting point in improving our documentation. The new layout (Figure 2) proved to be much more inviting and easy to read than the old. Consider this comparison:

Old	*New*
Two columns of justified text set in 9-point Univers. The small justified text was hard for some below-average readers to read, hard for our layout artists to correct, and hard for workers to read in a dimly lit factory.	One column of text, ragged right margin set in 11-point Schoolbook. These changes made the text easier for the reader and easier for our graphics department to work with.
The ½ inch page margins severely decreased the amount of white space on the page, making it uninviting and more difficult to read.	A 2½ inch left margin and limited line length to a maximum of 60 characters added helpful white space.
All headings were set in 9-point bold Univers. This gave all headings equal emphasis, thus making them all seem equally important. Located within the text columns, the heads were difficult to locate.	Major headings were set in 13-point bold Megaron and minor headings in 11-point bold Megaron, thus providing two levels of emphasis. Heads in the left margin make them easier to locate.
No color.	Chapter and section titles printed in blue. Coloring also made headings easier to spot.
Figures were often located several pages away from the text referring to them. Searching for figures after reading a figure reference was distracting and time-consuming.	Figures are dropped into the text where the references appear. This technique is especially important when showing someone what to enter into a computer and the display that should result.

We thought the more readable format provided an excellent starting point for our new approach, and the design was one that we could continually improve.

Document Scheduling

We were badly understaffed. Frequently, for example, we had to complete documentation projects in half the time normally alloted. Our writers had difficulty writing good documents under such severe time constraints. To help solve this problem we educated our marketing and engineering departments about the following: the value of giving us advanced notification on documentation needs so that we could become involved earlier in product development, the value of hiring additional technical writers to support our products, and the costs of not allowing enough time for documentation to be done properly (i.e., phone calls, complaints, accidents). By winning their cooperation, we have increased our writing staff and schedule projects more realistically.

Writing

Our writers had excellent technical backgrounds, but they lacked formal training in technical writing and exposure to the latest techniques for developing documents. To meet this need, we did two things. First, we hired professional and technical writers who had completed writing programs. Adding these writers created a more balanced department. These new writers provided a fresh outlook on documentation while our established writers provided our new writers with technical insights.

Second, we enlisted the help of Carnegie-Mellon University's Communication Design Center, specifically Erwin Steinberg and G. H. Jones to do the following:

Develop a training program for our writers

Develop a publication style manual containing guidelines on style and design

Begin an on-going objective evaluation of our documents

Create a seal of approval that could be printed in manuals that lived up to the high standards prescribed by the style manual. The seal would indicate our interest in serving our customers better through easy-to-understand documentation.

In addition to creating the style guide, Allen-Bradley and Carnegie-Mellon developed a document called "Writing Guidelines for Vendors" (Figure 3). We give these guidelines to vendors and consultants who produce documentation for us. Vendors are required to follow these instructions to have their documentation accepted. The guidelines discuss techniques such as writing in the active voice and using personal pronouns. By specifying up front the way we want the material handled, we ensure that we get a well-written document.

Document testing

We now test our documents before they reach the field. We use two types of tests: the user test and the customer test. In the user test we take manuals to the Customer Training Center and test them with representative users of our products. We have these users learn how to use a product by following the manual. We encourage them to tell us where they're having problems and what they're feeling. In this way, customers "edit" the manual. This test shows us where we've omitted information, been unclear, and misleading. In some cases we detect problems with the design of the product before it gets into the field.

In the customer tests we send our products to customers for evaluation. With these products we include a test copy of the manual and a questionnaire (Figure 4). The questionnaire asks the customers to comment on different aspects on the manual such as technical accuracy and ease of use. From the users' answers, we can determine how well the manual has supported a product that has actually been used in a factory environment.

By running these tests, we assure quality of our documents before the wider range of our customers gets them. In this way, we avoid upsetting our custom-

Figure 4. Customers evaluate Allen-Bradley manuals by completing this questionnaire.

Beta Site Manual Questionnaire

Introduction

Use this questionnaire to report on the quality of the manual.

1. How frequently did you have to flip back and forth between pages as you read this manual? (Place a check mark on the chart.)

|<----------|----------|----------|---------->|
Never — 4 times per chapter — 8 or more times per chapter

For which of the following reasons did you have to flip back and forth between pages?

____ To refresh my memory.

____ To find an illustration or table.

____ Because information about a topic was scattered throughout the text.

____ Because the headings were misleading.

____ Because the information was poorly sequenced.

Please list those places you recall having to flip back and forth.

____

____

____

____

2. How often did you have to re-read passages? (Place a check mark on the chart.)

|<----------|----------|----------|---------->|
Never — 4 times per chapter — 8 or more times per chapter

Figure 3. These "Writing Guidelines for Vendors" provide plain English Guidelines for consultants or vendors who produce documentation for Allen-Bradley Company.

Vendor Guidelines | Systems Division

Allen-Bradley Writing Style Guidelines for Vendors

How to Design and Write Manuals

Allen-Bradley's technical writers design and write performance-oriented manuals to serve their users' needs. They write manuals with the users in mind to show them how to work with the product, not merely how the product works. Basically, our writers accomplish this by:

- helping the users picture themselves in the text
- choosing words that the users can understand
- writing clear, straightforward sentences
- organizing the text for the users

Following is a list of twelve guidelines with examples. We suggest that you follow these guidelines when writing your documents for our users.

Help Users Picture Themselves in the Text

Guideline 1: Address the reader directly.

Original

It is suggested that the wire should be connected to the terminal by the engineer when the switch-box assembly is completed.

Revised

We suggest that you connect the wire to the terminal when you finish assembling the switch-box.

Original

With operator commands entered from the keyboard, an operator can display the program directly on the CRT.

Revised

By entering commands on the keyboard, you can display the program directly on the CRT.

ers with mistakes and omissions, and reduce the number of manuals we must reprint due to errors.

Persuading management

Our managers were well aware of the problems caused by inadequate documentation, but they did not always agree with the techniques we wanted to use for improvement. For example, they thought that using personal pronouns and the active voice in technical documents was improper because it removed that objective scientific tone associated with technical documents. To convince them plain English and effective design would work, we cited document design research that supported our techniques. With scientific backgrounds, they responded to the research tradition and began to accept our changes.

Now plain English and readable formats are the corporate standard for Allen-Bradley Company. In addition to pleasing our customers, the use of plain English and clear design is providing us with the following benefits:

Our phone center used to receive more than 50 calls about our products each day. We began shipping the first simplified manuals with a product over a year ago. Since then we have received an average of only two calls a month regarding the product and the manual.

It's hard to determine the amount of sales that were influenced directly by our documentation. Our sales are increasing. Our sales force is selling more systems because it can learn about them more quickly. Distributors report that the better the documentation, the less time they have to spend in the field teaching a customer about a product.

Because we are an international company, we must translate our manuals into Japanese, German, French, and many other languages. Clearer documents that use fewer words are easier for human translators to understand and less expensive for computers to translate (computer-assisted translations cost about 20 cents a word). In addition, a document printed in a foreign language such as Arabic or German requires 125% more space, so writing in plain English saves graphics time and space.

In the past two years technical writing at Allen-Bradley has come a long way, but implementing the plain English program wasn't easy. First, we had to convince management that plain English would benefit our company. Second, we had to teach our writers how to use it. Third, we had to teach our graphic designers how to present it. Now, plain English is selling itself as a cost-effective way to communicate.

A Program for Improving Documentation

Erwin R. Steinberg
Carnegie-Mellon University, Pittsburgh

The Communications Design Center of Carnegie-Mellon University undertook two projects for the Systems Division of Allen-Bradley: (1) A management study of the Allen-Bradley manual-writing process, which addressed problems with the content and presentation of information in manuals, training of technical writers, enforcement of document tracking and style guidelines, and needed critical documentation. (2) A Handbook for Manual Writers. The Center helps Allen-Bradley with quality control in the manual-writing process by assigning reviewers to sample sections of manuals in draft form, to comment on problems, and to suggest additions or revisions. Allen-Bradley manuals carry the seal of the Center to attest to the procedure established for drafting, testing, and editing those manuals (but not to the readability of any particular manual).

When the Communications Design Center of Carnegie-Mellon University agreed to develop a program to improve the manuals at the Systems Division of Allen-Bradley, those of us on the project spent considerable time trying to understand the division's documentation problems. We talked with the manager of commercial services, the department responsible for producing the manuals, and with the supervisors of the technical writers. We toured the plant and learned what products they made and how the products worked. We talked with people in engineering, marketing, and training. Finally, we read: advertising pieces, sales brochures, product data sheets, and manuals—many manuals.

We read the manuals with two purposes in mind: to prepare a two-day workshop for the technical writers who wrote the manuals and to plan a *Handbook for Manual Writers* that we had agreed to write. At this stage, "we" were two members of Carnegie-Mellon University's Communications Design Center who were also members of the university's Department of English. By the time we were ready to give the workshop, we had added to our team two graduate students enrolled in our program for the M. A. in Professional Writing. One of the graduate students had already completed a year in the program and had taken my course in professional and technical writing; the other had completed a semester. During the two-day workshop, the graduate students attended some sessions and spent some of the time familiarizing themselves with manual-writing problems by talking with Systems Division personnel.

Reprinted by permission of the publisher, *Visible Language,* Rhode Island School of Design, Providence, RI 02903.

A workshop for manual writers

We began the workshop with a discussion of the Shannon and Weaver "Schematic diagram of a general communication system"[1] and the Lasswell formula (who, says what, in which channel, to whom, with what effect)[2]; and we worked hard at the concept of audience. Then, using examples from the manuals, we addressed problems that we had found in them: the importance of tailoring the manual to the needs of the users; the importance of the introductory section of the manual and of the introduction to each section; organization; the way style and usage affect tone and comprehension, including such matters as personal pronouns, direct address, word choice, wordiness and redundancy, active and passive voice; consistency of language; sentence and paragraph structure; and layout and design. We also discussed methods by which the technical writers obtained their technical information and the manual-writing process, from assignment to final approval.

A management study

Before we could undertake writing the handbook, we were asked to do a management study of the manual-writing process. Our two graduate students spent several days at the Systems Division and then returned to write a report, the essence of which was:

I. Plain English and document design
  Problems with the content and presentation of information: Unclear language, insufficient examples, lack of glossary, lack of indexes, poorly designed table of contents, incomplete information, incorrect information, and lack of user-oriented logic in the organization of manuals.

II. Management
  Lack of proper training: Technical training for writers, training for all staff on principles of audience analysis, and training for writers on how to apply principles of audience analysis.
  Lack of enforcement of Allen-Bradley document tracking and style guidelines.
  Lack of critical documentation: A text introducing the fundamentals of programmable controllers, hardware compatibility charts, pocket guides for use on the plant floor, and case studies with step-by-step descriptions of normal system configurations.

A handbook for manual writers

We planned a handbook in five sections; planning, composing, designing, testing, and style and usage.

Chapter I. Planning the manual

1. Interviewing engineers, marketing and sales personnel, training center instructors, and customers
2. Learning to use the product
3. Analyzing the manual's prospective audience
4. Constructing a performance-oriented outline
5. Editing your first draft from the user's perspective
6. Conclusion

We adopted the terms "performance-oriented" and "topic-oriented" (the latter does not appear in the foregoing outline but does in the handbook) from the U.S. Army's *Guidebook for the Development of Army Training Literature*.[3] Here are the Army's definition of the two terms and accompanying examples:

> Topic-oriented writing focuses on the generalizations and concepts which constitute a body of knowledge — it tells "about" a subject area rather than telling "what to do" or "how to do it." (p. 6)
>
> **CBR: The Local Alarm**
>
> The individual suspecting or recognizing this attack will mask first and then give the alarm. The vocal alarm for chemical agent attack will be "SPRAY" for a spray attack, and "GAS" for an attack delivered by other means. The vocal warning is intended for those individuals in the immediate vicinity of the person recognizing the attack. The vocal alarm does not take the place of the sound alarm or the visual signal to alert a unit of a chemical attack. (p. 7)
>
> Performance-oriented writing focuses on the duties and tasks a user is expected to perform and the information he needs in order to perform these duties and tasks — it tells the user "what to do" and where possible, "how to do it." (p. 6)
>
> **CBR: The Local Alarm**
>
> How to Give the Local Alarm
>
> In case of Chemical attack, use these steps to give the local alarm:
>
> a. Put your mask on first;
> b. Give a vocal alarm — If spray attack, say "SPRAY"; for all other kinds, say "GAS";
> c. Give the sound alarm, visual signal, or both, as directed in your Unit SOP;
> d. Pass the warning to the Unit Commander as directed in your Unit SOP. (p. 9)

Chapter II. Composing the draft

1. Writing introductions
1.1. Writing general introductions
1.1.1. Purpose
1.1.2. Audience
1.1.3. Definitions of major terms
1.1.4. Related publications
1.1.5. Optional sections in the general introduction
1.2. Writing chapter introductions
2. Writing definitions
3. Writing descriptions
3.1. Physical descriptions
3.2. Process description for understanding
4. Writing procedures
4.1. How the correct choice makes a difference
4.2. Narrative format
4.3. Step-by-step list
4.4. Step-by-step visual aids
4.5. Action-response
4.5. Question list
4.7. Decision tables
4.8. Flowchart
4.9. Narrative Flowchart
5. Writing examples
5.1. Extended examples
5.2. Short examples
6. Conclusion

Chapter III. Designing the layout and graphics

1. Allen-Bradley Design Standards
2. Headings
3. Layout and use of white space
4. Graphics
4.1. Labels
4.2. Illustrations
4.3. Diagrams
4.4. Tables
4.5. Typographic highlights

We need not discuss Chapter IV, "Testing by user edits," because Barry Jereb has covered it in his section "Document testing." In Chapter V we dealt with the kinds of matters we considered in the workshop, discussed early in this paper, and in the vendor guidelines, a discussion of which follows.

Style guidelines for vendors

As Jereb has already explained, vendors and consultants produce documentation for the Systems Division which frequently must be integrated with the division's documentation. Even if it is used separately, Allen-Bradley wants to be sure that it is comprehensible. We produced, therefore, "Writing Style Guidelines for Vendors," the introduction and outline for which follows:

> Allen-Bradley's technical writers design and write performance-oriented manuals to serve their users' needs. They write manuals with the user in mind to show them how to work with the product, not merely how the product works. Basically, Allen-Bradley writers compose performance-oriented manuals by: helping the users picture themselves in the text, choosing words that the users can understand, writing clear, straightforward sentences, and organizing the text for the user.
>
> Following is a list of twelve guidelines with examples. We suggest that you follow these guidelines to write your documents for our users.
>
> *Helping users picture themselves in the text*
>
> 1. Address the reader directly, by name, or by using a pronoun.
> 2. Write in the active voice.
> 3. Use action verbs rather than nouns made out of verbs when you can.
>
> *Choosing words that the users can understand*
>
> 4. Choose your words with care. Avoid jargon. Define or explain technical terms that you cannot change.
> 5. Don't rename for the sake of variety.
> 6. Don't use extra words.
>
> *Writing clear, easy-to-follow sentences*
>
> 7. Put the parts of each sentence into logical order.
> 8. Untangle convoluted sentences.
> 9. Use lists when you have several items to discuss. Check that all items in the list have similar grammatical constructions.
> 10. Rewrite multiple negatives as positive sentences when you can.
> 11. Avoid noun strings.
>
> *Organizing the text for the user*
>
> 12. Think of your user when organizing text.

Accompanying the first eleven guidelines are examples: first, poorly worded originals, and then suggested revisions. The last guideline reads:

> Think of your users when organizing text. Understand the needs of your users so you can position the users correctly in the text; describe how they use the product, not merely how the product works. Lay out and design information so users can read and use it easily. Provide meaningful headings so users can find information easily. Summarize information when appropriate so users can review important information.

And the guidelines conclude:

> Writing for users isn't easy. You need to think about what the users want to know and what they must know to do their job well. Think of what job the users will do, not what the product will do. This will help you gain a performance-orientation to writing. By following these guidelines, you will begin to compose manuals for the users. For more suggestions on how to write performance-oriented manuals directed to your users needs, ask to see Allen-Bradley's *Publication Style Guide*.

The technical writing supervisors are currently putting the finishing touches on the handbook to make sure it speaks to all the needs of the manual writers. When they have finished, Allen-Bradley will print the handbook and the supervisors will distribute it to their technical writers, run a workshop to instruct the writers in how to use it, and monitor the writing to be sure that the writers do use it.

Quality control

To assist the Systems Division with quality control in the preparation of manuals, the Carnegie-Mellon Communications Design Center has been included in the manual-writing "loop." Graduate students sample sections of the drafts of new manuals and comment on any problems they find or suggest additions or revisions. The following form, which a graduate student fills out and attaches to each draft after sampling that draft, shows the kinds of things a reviewer looks for:

CDC form for evaluation of technical manuals

	Revision required		
Area reviewed	*Much*	*Some*	*None*
Definition of audience			
For manual	______	______	______
For chapter	______	______	______
Content			
Table of contents	______	______	______
Definition of caution, warnings, notes	______	______	______
Glossary	______	______	______
Index	______	______	______
Completeness	______	______	______
Accuracy	______	______	______

Organization			
Usage and style			
Passive voice			
Wordiness			
Undefined terms			
Noun strings			
Parallelism			
Consistency			
Jargon			
Punctuation			
Layout and graphics			
Graphics with text			
Graphic labels			
Usefulness of graphics			
Production			

The Seal of the Communications Design Center

To testify to the procedure, each new manual produced by Allen-Bradley's Systems Division will carry the seal of the Communications Design Center and the following statement:

> This seal means better documentation for you because this document has been written and tested according to guidelines developed by the Communications Design Center (CDC) of Carnegie-Mellon University (CMU) specifically for Allen-Bradley.
>
> The CDC periodically evaluates A-B's application of these guidelines; however, Allen-Bradley warrants that neither the CDC nor CMU nor any unit of the university in any way certifies the technical accuracy of the document. The CDC seal indicates solely that the CDC has approved the planning and testing of the organization, layout, and writing style of manuals and is evidence of Allen-Bradley's interest in serving its customers better through easy-to-understand documentation.

The seal thus attests to the procedure and not to the readability of any particular manual.

1. Claude E. Shannon and Warren Weaver, *The Mathematical Theory of Communication* (Urbana: University of Illinois Press, 1949), p. 5.

2. Harold D. Lasswell, "The Structure and Function of Communication in Society," in *The Communication of Ideas,* ed. Lyman Bryson (New York: Harper, 1948), p. 3.

3. Richard P. Kern, Thomas G. Sticht, Diana Welty, and Robert M. Hauke, *Guidebook for the Development of Army Training Literature* (Springfield, Va.: National Technical Information Service, U. S. Department of Commerce, 1975)

Why Government Has Difficulty Communicating

Irene A. Etzkorn
Siegel & Gale, New York

Government and paperwork are closely related kin. Yet thousands of forms, applications, letters, and notices written by government agencies result in little communication. One of the most recent and well-publicized examples of this phenomenon was a new W-4 Withholding Form introduced by the Internal Revenue Service in 1987. Responding to the mass confusion that the form created, the IRS introduced a second "easier" version mid-year. When its second attempt also failed, the IRS conceded that taxpayers who had too little tax withheld should not be penalized because of the complexity of the form. A rare event? On the contrary, as this article will show, similar communications gaps occur frequently between governments and their citizens.

Who's to blame—the writers or the readers? Do government agencies face a unique communications problem? What can be done to lessen the paperwork burden? In this article, I discuss these questions and present case histories of several successful simplification projects.

WHY GOVERNMENT HAS DIFFICULTY COMMUNICATING

For those who relish irony, the need for clear communication in government is a particularly delightful example. The unique responsibility of government to avail its citizens of their rights and obligations as well as its duty to tackle society's most fundamental and complex problems makes clear communication critical.

In fact, it is possible that clear communication is most important in government, where the course of history is determined by mutual understanding. Yet, ironically, the very nature of government—a collective entity with many voices—undermines its ability to communicate. Government communicates in many ways, including letters, notices, pamphlets, forms, and applications.

It is especially difficult for government to communicate, however, for a variety of reasons:

- *The vastness and diversity of government's audience.* No corporation has an audience as broad as all taxpayers, all citizens, or all retirees. The diversity of the audience makes it difficult to determine a proper tone and level of communication.
- *Constantly changing laws.* To remain accurate and up-to-date, government documents must constantly be revised and patched. And as we all know, patchworks are characterized by loose threads and inconsistent patterns.

- *A misguided belief that government writing must sound "official."* Too often, "official" sounds pompous and convoluted.
- *A tendency to write to protect rather than to communicate clearly.* Documents that are written to serve as records or legal proof often bury requests for action and important instructions in a maze of factual detail.
- *The need to give all citizens equal access to information.* In an attempt to be comprehensive and to anticipate the needs of all its citizens, writers of government documents often unintentionally make their readers victims of "information overload."
- *A tendency to practice "false economies."* For example, the Internal Revenue Service counts the number of lines on a form to measure its complexity. This results in forms with multipart lines (8 a,b,c,d) rather than individual line numbers (8, 9, 10, 11), because shorter is easier. Not in this case.

In fact, multipart line numbers complicate computation because it is easy to overlook or skip parts of lines when adding them up.

WHY GOVERNMENT DOCUMENTS MUST BE SIMPLIFIED

What is the result when government fails to communicate clearly? At least, frustration. At most, injustice. And always, excessive cost to its citizens.

In the United States, for example, citizens who fail to provide certain information to the Federal government may be penalized (census and income tax information are two notable examples). This "penalty-carrying" information constitutes 42% of the U.S. Federal paperwork burden.[1] Yet citizens who attempt to comply with the law are often frustrated by confusing, lengthy, or redundant paperwork.

The level of frustration can be best appreciated by noting specific examples of documents received by the public. Consider this excerpt from a letter sent to a Medicare claimant:

> Our Medicare Department has reviewed these claims and have [sic] determined that no additional allowances are warranted. They were paid correctly to the doctors' new and old profiles.[2]

This letter triggered a class action lawsuit. The person who received it—an elderly man covered by government Medicare insurance—sought to determine why he received different amounts when he submitted a series of identical claims for his wife's cancer treatments.

Clearly this letter did not adequately answer his question, although the letter ended with the closing, "We trust we have clarified the matter for you." The federal judge who heard the case was not satisfied with the answer either and ordered gobbledygook removed from Medicare forms: "The language used is bureaucratic gobbledygook, jargon, double-talk, a form of officialese, federalese and insurancese and doublespeak. . . . It does not qualify as English."[3]

This excerpt from a Department of Energy Request for Proposals is yet another example of "governmentese."

> Technical evaluation criteria are listed below in descending order of importance. The relative importance of Criterion 1 is approximately 1½ times as important as Criterion 2. Criteria 2 and 3 are of equal importance; combined they are four times as important as Criterion 5. Criterion 5 is approximately ⅓ as important as Criterion 1. Subcriteria are listed in generally descending order of importance.[4]

WHAT GOVERNMENT IS DOING TO REDUCE PAPERWORK BURDEN

A few government agencies have chosen "to light a candle rather than curse the darkness." Citizens' increasing clamor for the "right to know"—as evidenced by the consumer movement—has not gone unheeded. In the United States, support for the plain English movement has grown considerably since its first real stirrings in the mid-1970s. By then, the consumer movement had gained momentum, and consumers demanded more information, greater rights, and self-help remedies. Market research studies indicated an increase in public participation and a growth of participatory democracy, a conviction that with proper information people would make informed decisions about the courses of their lives. Support grew at the state level, and early efforts focused on writing consumer contracts in plain English (primarily banking and insurance documents).

The signing of the Paperwork Reduction Act of 1981 marked a turning point in the United States, when the federal government acknowledged the "paperwork burden" imposed on its citizens and instructed federal agencies to reduce paperwork and eliminate gobbledygook from regulations.

Some members of the executive branch have made plain English personal crusades. Most notably, when Malcolm Baldrige became secretary of commerce in 1980, he decreed an end to obfuscation in the department. Recalling he "literally could not tell whether he was saying yes or no" in many of the letters being drafted for his signature, he called a meeting of department letter writers and ordered, "Write directly and clearly . . . and always say yes or no." He also had the department's word processors reprogrammed to reject words such as "optimize, interface, prioritize, and finalize."[5]

Perhaps the most far-reaching organization of plain English advocates is in Canada. The Canadian Law Information Council was established ten years ago to improve the quality and timely delivery of legal information to lawyers, law schools, judges, and the public. The Council acts as a national clearinghouse, a resource center, and a research organization. A project to study documents from Canadian small claims courts is currently in progress. The study will look at public access to information, the relevance of available information, and its readability and visual effectiveness.

The movement for plain language is a worldwide concern and is not confined to English. Efforts to write public documents in language that ordinary people can understand are underway in the Netherlands, Sweden, Nigeria, Romania, and Ireland.

These government efforts have produced many significant accomplishments. For example, between 1980 and 1983, the U.S. reduced paperwork government-wide by 32%, far surpassing its 25% goal.[6] This means that over *400 million hours* of paperwork have been eliminated since 1981 by changing filing requirements and eligibility rules so that more people can use shorter or fewer forms.

On a lesser scale, the U.S. Federal Communications Commission used to have five full-time employees who spent all their time answering telephone calls about the citizens band radio rules. Then, in 1977, the FCC reorganized and rewrote the rules in plain English. The phone calls soon stopped, and all five employees were subsequently transferred to other jobs.

SHORTCOMINGS OF EXISTING GOVERNMENT EFFORTS

If government agencies have undertaken simplification projects, why hasn't greater improvement been realized? In general, government simplification efforts can be characterized by a number of shortcomings:

a) Government's "mechanistic approach" to simplification.
   Legislation which mandates the use of readability formulas is a typical symptom of the mechanistic approach.

b) The use of inappropriate measures for calculating the public's "paperwork burden."
   The current formula for estimating the burden imposed by IRS tax forms only measures the time to actually fill out the forms and does not take into account time taxpayers spend keeping records and obtaining information, nor the time and money spent for processing incomplete forms and correcting errors.

c) The focus, particularly in the United States, on reduction rather than simplification.
   Reducing volume without simplifying the content of the remaining forms does not adequately address such issues as the time required to fill out a form, to read and understand the instructions, and to develop, compile, and review the information collected. Changing filing requirements and eligibility rules is only one aspect of simplification.
   In addition to reducing the amount of information collected and the quantity of paperwork required to elicit such data, government must make sure that the essential paperwork is written in clear, readable language appropriate for its intended audience, and formatted in a manner that is suitable for its use.

d) Incremental rather than comprehensive change.
   Government agencies tend to phase in changes rather than introduce an entire document or system of simplified documents. Unfortunately, this patchwork approach adds to the overall complexity because the public

must get used to a slightly different form or document each time it is issued.

Problems also arise when changes, such as the use of graphics, are introduced without a real understanding of the simplification process. The result can be "technicolor confusion."

e) Narrow definition of communications.
There is often immediate acceptance of the need to simplify forms, but a lack of carry-over to all forms of communications—letters, leaflets, guides, and manuals.

The situation, however, is not hopeless. There have been several successful government simplification projects.

IRS TAX FORMS SIMPLIFICATION PROJECT

Perhaps the most significant government-sponsored simplification effort was undertaken in 1981, when the U.S. Congress mandated that the Internal Revenue Service (IRS) simplify the entire system of federal income tax forms and instructions.

From 1979 to 1982, Siegel & Gale, a private consulting firm specializing in document design, was retained as the primary contractor to simplify the IRS's system of personal income tax forms. The project called for a complete restructuring of these forms, rather than cosmetic changes in language and design.

Siegel & Gale's work led directly to many substantial improvements in the current system:

- taxpayer aids, including work charts, examples, easy reference charts, and tax tips that have been adopted by the IRS;
- introduction of the 1040EZ;
- simplified schedules for itemized deductions, income averaging, and child care credits; and
- the testing procedure used to evaluate and refine the prototype forms, which has become an integral part of the IRS's forms development program.

Overall, the prototype tax forms were highly successful when judged by a number of measures:

- reduced error rates;
- shorter completion time; and
- more positive taxpayer attitudes.

REVENU MINISTRY OF QUEBEC TAX FORMS SIMPLIFICATION

In 1982, under the leadership of a new minister, Revenu Québec undertook an extensive program to humanize its forms and services. The department

Line 2. Wages, salary, tips (continued)

You should receive your copy from your employer by February 1. If you don't receive your statement by then, ask your employer for it. If you still haven't received your statement by February 15, contact your local IRS office, Whether or not you get a wage statement, you must still report your earnings from that job.

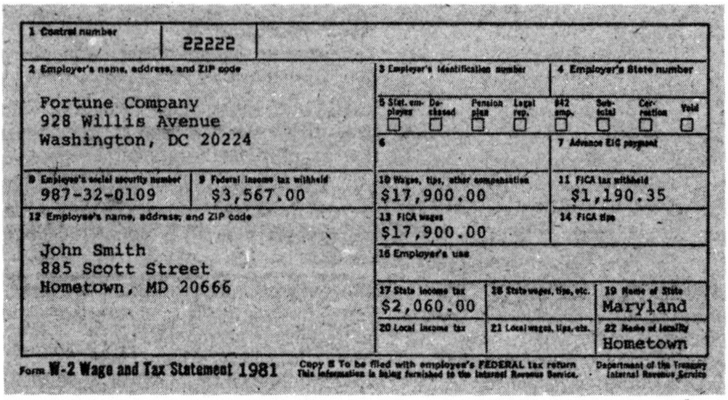
1 Control number 22222
2 Employer's name, address, and ZIP code
Fortune Company
928 Willis Avenue
Washington, DC 20224
3 Employer's identification number
4 Employer's State number
5 Stat. employee Deceased Pension plan Legal rep. 942 emp. Subtotal Correction Void
6
7 Advance EIC payment
8 Employee's social security number 987-32-0109
9 Federal income tax withheld $3,567.00
10 Wages, tips, other compensation $17,900.00
11 FICA tax withheld $1,190.35
12 Employee's name, address, and ZIP code
John Smith
885 Scott Street
Hometown, MD 20666
13 FICA wages $17,900.00
14 FICA tips
16 Employer's use
17 State income tax $2,060.00
18 State wages, tips, etc.
19 Name of State Maryland
20 Local income tax
21 Local wages, tips, etc.
22 Name of locality Hometown
Form W-2 Wage and Tax Statement 1981
Copy B To be filed with employee's FEDERAL tax return
This information is being furnished to the Internal Revenue Service.
Department of the Treasury Internal Revenue Service

Model 1981 wage statement, Form W-2, showing wages and amounts withheld for Federal, state, and Social Security taxes.

Work Chart for Wages, Salary, Tips

If you received income from more than one job in 1981, you should receive a wage statement from each of your employers. You may find it helpful to use the work chart below to figure your total earnings from jobs.

***Tax Tip.** If you held more than one job in 1981 and earned more than $29,700, too much Social Security tax may have been withheld in 1981 (see page 23).*

Employer	Wages, salary, tips
	$.
	+$.
	+$.
	+$.
	+$.
	+$.
	+$.
Total	=$.

Write the total on line 2 of your tax return.

Self-employment. If you had any income from self-employment, you cannot report that income on the Short Form. You must use Form 1040 instead (see page 3).

Military or clergy. If you were in the military or were a member of clergy, special rules apply to your income. For more information, see IRS Publication 17, Your Federal Income Tax.

New for 1981

Community property states. New rules apply to some taxpayers who live in community property states. These states are Arizona, California, Idaho, Louisiana, Nevada, New Mexico, Texas, and Washington. Beginning in 1981, you do *not* have to include on line 2 any amounts that your spouse earned from employment if **all four** of the following apply:

1. You lived apart from your spouse for all of 1981, and
2. You do not file a joint return, and
3. You do not give any part of your income from employment to your spouse, and
4. Your spouse does not give any part of his or her income from employment to you.

For more information, see IRS Publication 555, Community Property and the Federal Income Tax.

Figure 1. A page from the simplified 1981 U.S. short form taxpayer guide for single taxpayers.

General instructions

This Guide contains a 1982 Short Form from Revenu Québec and instructions.

This Short Form was sent to you based on the return you filed last year. Since the Short Form can be used to report most types of income and deductions, a majority of taxpayers can use it. However, if you have any of the types of income or deductions listed on page 2, you may *not* use the Short Form. This Guide will help you determine the form you need.

To complete your return quickly and accurately, follow these steps:

- **first,** be sure you must file a Québec tax return.
- **second,** be sure you can file the Short Form.
- **third,** collect your tax records.
- **fourth,** prepare and mail your tax return.

First, be sure you must file a Québec tax return

If you resided, worked or conducted business in Québec during 1982, you usually have to file a Québec tax return. To be sure you need to file, read the residency and income requirements described below.

Residency requirements
You *must* file a Québec tax return if any *one* of the following applies:
your legal residence was in Québec on December 31, 1982 (even if you were employed in another province or country).
or you conducted business in Québec during 1982 (even if you resided in another province or country or emigrated from Canada).
or you were employed in Québec during or prior to 1982 and resided outside Canada.
or you sold taxable Québec property during 1982 and resided outside Canada.
or you were residing in Québec on the day you emigrated from Canada.

Working but not residing in Québec. If you were employed in Québec during 1982, but on December 31, 1982 resided in another province of Canada, you are *not* required to file a Québec tax return. Attach a copy of your Slip 1, Revenus d'emploi, to your Federal income tax return and Canada Revenue will credit you for the amount of Québec tax deducted. See also the section called *General averaging* on page 14.

Income requirements
To see if you need to file a Québec income tax return, refer to the table below.

If, as of December 31, 1982, your marital situation was:	You must file a Québec tax return if:
Single, and you were under age 65	Your total income was over 4 780$
Single, and you were age 65 or older	Your total income was over 6 730$
Married, and you and your spouse were under 65 and your spouse did *not* have any income in 1982	Your total income was over 8 290$
Married, and you were over 65 and your spouse did *not* have any income in 1982	Your total income was over 10 240$

Figure 2. A page from the simplified general instructions of the 1982 Revenu Québec short form taxpayer instruction guide.

retained Siegel & Gale to redesign and rewrite the individual tax forms and instructions as the leading edge of this program.

Siegel & Gale developed prototype versions that combined an assortment of tax forms, instructions, special tax schedules, and government allowances. The system included long and short form versions in both French and English, and redesigned tax records (similar to our W-2s) to further simplify tax preparation.

The entire system took nine months to develop, test, and refine, and was implemented for tax year 1983. The new forms were extremely well received by taxpayers and tax professionals alike and produced immediate cost savings in processing.

MODEL CENSUS FORM

During the 1970s, the U.S. Bureau of the Census conducted several nationwide pretests in preparation for the 29th Decennial Census of April, 1980. In these pretests, the Bureau replaced census takers, or "enumerators," with self-contained questionnaires. However, many of the questionnaires were completed incorrectly, resulting in costly follow-up visits by enumerators.

National studies had also concluded that the census form's appearance evoked "unnecessary anger, hostility and resistance" and recommended that the questionnaire be rewritten and redesigned to be clear and attractive to the respondent. The Bureau of the Census retained Siegel & Gale to develop a people-oriented model census form that incorporated state-of-the-art computer techniques.

The content of the existing census form was analyzed, and the information needed by the bureau was determined. Computer experts were consulted in order to provide computer formats and to identify ways to eliminate duplication and simplify data collection through computerized checks and matches. The instructions were incorporated directly onto the questionnaire, and a clear question-and-answer format was used. In addition, the questionnaire was redesigned using an attractive format to encourage respondents to complete the form. To improve readability, a larger, more legible typeface was used, and the questions were arranged in distinct columns.

FOOD STAMP FORMS FOR THE U.S. DEPARTMENT OF AGRICULTURE

New federal regulations required the development of a single set of food stamp forms that could be used nationwide. These had to be simply written, clearly designed forms oriented to the needs of the public, rather than the administrators. Siegel & Gale was retained to develop new forms that would simplify the application process and make it easier for recipients to understand changes in their status.

Applicants, caseworkers, and regulators were interviewed, and language aimed at a sixth-grade reading level was used because many applicants and

U.S. Food Stamp Program
Notice of Eligibility

Case number ________________

Receiving Your Food Stamps

Date ________________________________

Name ________________________________

Address ________________________________

____

Dear ________________________________

We're writing to tell you about your food stamp benefits. The information you should know is next to the box marked below:

☐ You'll receive food stamp benefits during the month(s) of:

________________________________.

The first month you'll receive ________________ which covers the month(s) of: ________________

________________________________.

After this first month you'll receive ________________ ________________ each month.

☐ Your food stamp benefits have gone up to $________________ a month because ________________

________________________________.

☐ Because you needed food stamps benefits right away, we postponed asking you to give us certain information. We now need you to bring or mail in the following information:

____

________________________________.

You may ask someone else to bring it in. If we receive this information by ________________, you won't have to reapply. You'll keep on receiving food stamps without a break if you still qualify.

If there's a change in the number of people in your household—or a change in your household's income—please get in touch with us. We have a Change Report Form that makes it easy to report these changes. You're responsible for keeping us informed about them.

If you don't think you're receiving the correct amount of benefits, you can have a fair hearing of your case. At the hearing you'll have a chance to explain why you think we made a mistake, and a hearing officer will decide if you're right. To request a fair hearing, just call me at ________________, write me, or fill out and return the form below. You can also call me if you want to know more about how a fair hearing works. For free legal advice, call ________________.

Sincerely,

- -

If you want a fair hearing, fill out this form, tear it off, and mail to:

____
Case number

____
Case worker

____
Name of person appealing

____
Your signature

____
Address

Telephone number

____

If you'd like to tell us why you want a fair hearing, use this space to give us the reason.

____

____

Figure 3. A simplified U.S. food stamp form prepared by Siegel & Gale.

recipients have little formal education. A letter format was created to inform recipients of changes in their status, and a tear-off portion was developed which could be mailed back in order to request a hearing. To "humanize" the forms, two different colors of ink were used to make clear distinctions between the various parts of the forms and to highlight the instructions.

Nine prototype forms were tested in five locations around the country. Comments were gathered from food stamp officers, consumer groups, and the business community, and revisions were made to the prototypes based on this field testing.

COST BENEFITS

A strong paperwork simplification policy offers immediate cost benefits as well as long-term social benefits.

Fewer, simpler forms and applications offer immediate savings in printing, storage, and photocopying. Less obvious, but more dramatic savings are realized from fewer completion errors, shorter training time for employees, and prompter, more accurate processing. According to the *Wall Street Journal*, March 21, 1983, "For every $1 spent on forms, $20 more goes for clerical costs to process them."

The potential social benefits of simplification arise from the unique responsibility of government to avail citizens of their rights and obligations. Paper, in the form of Medicare forms, Social Security notices, income tax returns, census forms, and food stamp applications, is the embodiment of government for many citizens. A government perceived as a clear, forthright communicator stands to gain respect and allegiance from its citizens.

SUMMARY

Conducting paperwork audits, rewriting reports, and redesigning forms will take time and effort, and in some cases, can be facilitated by the expertise of outside consultants. Similarly, setting standards for measuring degrees of simplification and demonstrating results requires some research and considerable effort. Methods for establishing, communicating, and maintaining design and language standards also must be determined to ensure the success of a simplification policy. However, none of these obstacles is insurmountable.

The tremendous, ongoing benefits to be realized from a comprehensive paperwork simplification policy far outweigh the expense and effort needed to implement the policy.

1. "Information Collection Budget of the United States Government," Fiscal Year 1984, Executive Office of the President, Office of Management and Budget (figure 3), p. 6.
2. Letter to Joseph David explaining why he was denied additional benefits.

3. Opinion of a U.S. federal judge, July, 1984.
4. Request for proposals issued by the U.S. Department of Energy's Western Area Power Administration, April, 1984.
5. Baldrige, Malcom, U.S. Secretary of Commerce. "Plain English in Business and Government," January 5, 1983, address at the Forum on the Productivity of Plain English sponsored by the U.S. Department of Commerce.
6. "Managing Federal Information Resources," Third Annual Report under the Paperwork Reduction Act of 1980, Office of Management and Budget, June, 1984, p. 8.

Expanding the Application of Simplified Communication: Helping People Use and Choose Products

Michael J. Floreak, Ronald G. Amodeo, and Lorraine K. Wilkin
Fitch RichardsonSmith, Worthington, Ohio

The value of simplified communication is evident in instructional documents, like bank forms and computer manuals. Yet it represents only one of the many potential interfaces between people and products. Our work focuses on identifying and simplifying the appropriate interface between people and products by simplifying communication about the product, the product itself, or some combination of the two. Our work also addresses using simplified information both to help people use products (the traditional domain of document design) and choose products (a new application).

INTRODUCTION

The focus of simplified communication has long been on documents that help people use products. In fact, it seems at times that the only documents ever discussed are the endless variations on "how to" manuals for computers. While these kinds of user documents do indeed need simplification and provide a welcome starting point for a simplification movement, there are other document forms that need attention as well.

In our consulting work, we've broadened our definition of document to include any interface point between people and products. This definition, of course, includes instructional manuals and forms systems—the traditional domain of document design. But we also suggest the possibility of some very different kinds of documents. We have, for example, simplified "documents" such as the interior of a travel agency, packages for educational toys, and a kit for cleaning chemical spills. We find some very exciting implications in these nontraditional document forms.

Our expanded definition of documents lets us look at simplifying how things work as well as how they read. Our work often involves redefining a product, rather than simply documenting it. And it permits us to turn some attention to the long-ignored process of helping people choose products. While document design has traditionally focused on the way people *use* products, the reality of the marketplace suggests a need for simplifying the way people *choose* products as well. As members of a consumer society, we're bombarded constantly with information on what and how to buy. The sheer volume suggests a need for making information simple and understandable.

An understanding of the different decision-making processes involved in using and choosing products, plus some fundamental principles of simplified communication (understanding the audience, identifying the primary purpose and defining the appropriate method), are the keys to developing information that supports these different decision-making processes.

In the remainder of this paper, we describe a metaphor for the using/choosing process and review the application of simplified communication through client projects involving nontraditional document forms: for Mallinckrodt, a communication product designed for industrial use; and for Carol Cable, a point-of-purchase information system for selling a consumer product.

CHUTES AND LADDERS: A METAPHOR FOR DECISION MAKING

In a decision-making process, people move through information much as they might physically move through space. They begin from a starting point, make a series of steps and reach an ending point. How they move through that space depends on their situation: where they start, where they want to end, how fast they want to move, and what stands between the two points.

Chutes and ladders seems an apt metaphor for describing the differences in how people move through information in the different processes of choosing and using products. In both processes, people begin from the starting point of a felt need. From that point, they move through information by making a series of decisions until they reach a desired end—satisfaction of their needs. How they progress through the information is related to the differences in the situations.

Chutes

When choosing products, people often look for shortcuts through information. They want to get where they're going by big jumps, eliminating any steps that will slow them down. They look for information that interests them and skip over information that doesn't interest them. The amount and kinds of shortcuts people look for are affected by the importance of the decision, the amount of background information they have, and the qualities of the product they are deciding on. Furthermore, because the feelings of need and satisfaction of need are very specific to individuals, different people require different kinds of information in order to choose a product.

Information that supports a choosing process, then, should address these vastly different needs and tastes to provide many potential beginning and ending points. The information should be presented in discrete, easy-to-process units that suggest a hierarchy of importance, but accommodate individual differences in understanding and need to allow quick movement from any one place to any other.

Ladders

When using products, people move through information in a more linear set of steps. They still want to move through the information as quickly as possible, and indeed may choose to skip some steps or slide down two steps at a time. However, their freedom of movement is directly tied to the task and the product being used. They need to perform certain invariant actions or follow a certain series of steps that are imposed by the product. Therefore, the product often suggests one path that's more efficient than others for most every user.

Information that supports a using process should provide steps through the process, appropriately spaced and marked. The information should guide people through at a comfortable pace—not force them to take dangerously large steps or move so slowly that it becomes tedious. It should also provide the option of proceeding cautiously or jumping off from a comfortable point. The information should be presented in units that are hierarchically structured and easy-to-process; and it should give people a strong sense of where they are in the process, letting them know what has come before and what comes after.

In certain ways, choosing and using products are very similar: they both involve moving people through information as efficiently as possible. However, the differences are also quite important because they suggest ways in which information supporting using and choosing processes need to be different. The case studies that follow further demonstrate these differences.

CASE STUDY: MALLINCKRODT

Simplified information that helps people use a product

Mallinckrodt, a large manufacturer and distributor of laboratory and electronic chemicals, came to us with a communication problem. Although an experienced company, established more than 100 years ago, Mallinckrodt's image was not as strong as it had once been. Feeling the pressure of competitors who had developed contemporary corporate images, Mallinckrodt wanted us to help them redefine their position in the marketplace and communicate an image that more clearly reflected the Mallinckrodt of today.

We suggested that providing simple, understandable information and products could be the key to differentiating themselves from the competition and reflecting their understanding of their customers and the marketplace. The opportunity to introduce this new commitment to simplified communication came with a new product.

Developing a product that's easy to use

Mallinckrodt identified a need within the chemical industry for a laboratory-sized kit to clean chemical spills. There were already industrial-sized spill kits on the market designed to handle spills of 4 to 24 liters. But since the kits were large and cumbersome, they did little to satisfy the needs of laboratory chemists who often work with volumes of one liter or less. Although there was an obvious need for a laboratory kit, only one other manufacturer offered one. Mallinckrodt seized a unique positioning opportunity by taking the laboratory-sized kit one step farther—making it easy to use.

We helped Mallinckrodt identify three potential audiences for this product: laboratory chemists, other laboratory employees who could come in contact with chemicals (for example, custodians), and the laboratory chemical market. Laboratory chemists were used to making do with the larger-sized spill kits. Those chemists fortunate enough to have access to the only laboratory-sized spill kit were forced to wade through a 30-page user manual to learn the specific cleanup processes for different types of chemical spills. Because chemists are like most other people, they're likely to wait until a spill has occurred before they begin investigating how to clean it up. And by then, the liter of hydrochloric acid they spilled would probably be penetrating the soles of their shoes and the vapors impairing their eyesight. Granted, a laboratory chemist probably has a general idea of what it takes to absorb and neutralize a chemical spill. But we were concerned with making it as easy as possible for them to find and use the absorbant and neutralizer they need in this emergency situation.

And what about the nonchemists? If a custodian sweeping the lab accidentally knocks over a bottle of acid, he can't just swab it up with paper towels. How would the custodian know what cleanup materials to look for? And if he finds the 30-page manual, will he be able to locate the correct cleanup procedure? We wanted to design a kit that is immediately obvious to everyone in or around the lab and contains everything needed to clean a 1 liter spill—including goggles, gloves, brush, scoop, waste bags—and easy-to-understand, illustrated instructions.

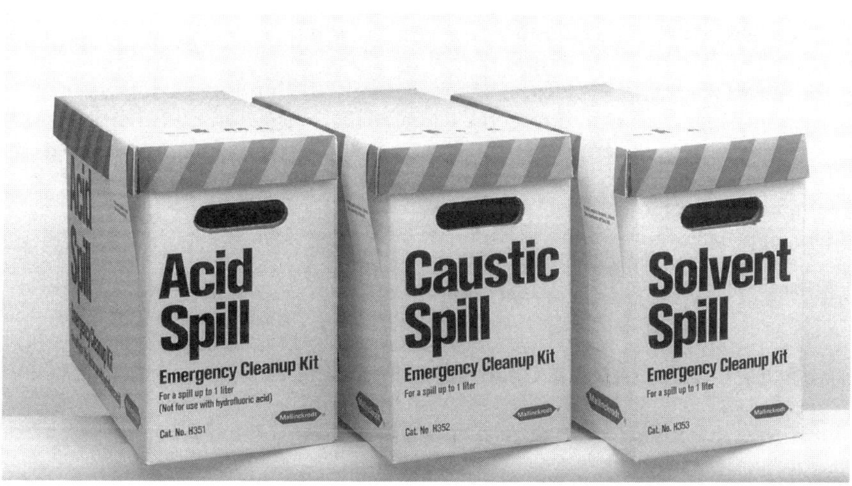

Figure 1. Bold headings on the packages identify their use even from a distance.

We also wanted this new product to address the needs of the third audience, the laboratory chemical market, by informing them of the new Mallinckrodt image. We wanted to reflect Mallinckrodt's true image through a well thought-out product that was as easy to use as it was to sell.

Rethinking traditional "how-to" information

Our approach to the spill kit product involved challenging traditional presentations of "how-to" information. Rather than treating the information as an add-on to the product (as a 30-page manual, for example), our goal was to make the product easy to use by redesigning the product itself. This approach dramatically affects how the product is used and perceived by customers.

The bold typographic presentation on the front of the carton is simple, yet visible from across the lab. It communicates clearly the type of spill the kit is designed to clean. And its simplicity becomes the foundation for a sharper, bolder, more direct image for Mallinckrodt.

The structure of the kit was designed not only to hold the kit components, but also to communicate information about how to use the product. For example, the safety seal across the top of the kit shows that the kit contents are intact and serves as a warning that, when broken, components of the kit may be missing.

Once the kit is opened, the user sees simplified instructions on the inside of the lid. Instead of providing a 30-page manual detailing the cleanup process for several types of chemicals, the entire cleanup process for a particular chemical spill is represented in a ladder of four meaningful, action-oriented steps: (1) Protect, (2) Absorb, (3) Neutralize, and (4) Cleanup. Each step is then followed with more detailed instructions for use.

In addition, each component of the kit is numbered, labeled, and organized in the order of use: (1) Protectors, (2) Absorbent, (3) Neutralizer, (4) Cleanup. The user simply locates and uses the components in the designated order. And just in case the user missed the instructions on the inside lid or carried the component away from the kit, the instructions for each component are repeated on the component itself.

This approach represents a significant shift away from a separate kit and manual. Here, the kit is the manual. The development involved a major rethinking of how the product works and how it is presented to customers.

CASE STUDY: CAROL CABLE

Simplified information that helps people choose a product

Carol Cable produces extension cords—thin and colorful ones for inside the house, thick and sturdy ones for the lawn, garage, or workshop. They offer the same product line as their competition, products of equal quality and durability, products of equal innovation. Recognizing the pressure of younger rivals who were producing colorful new packages and expanding product lines with all-

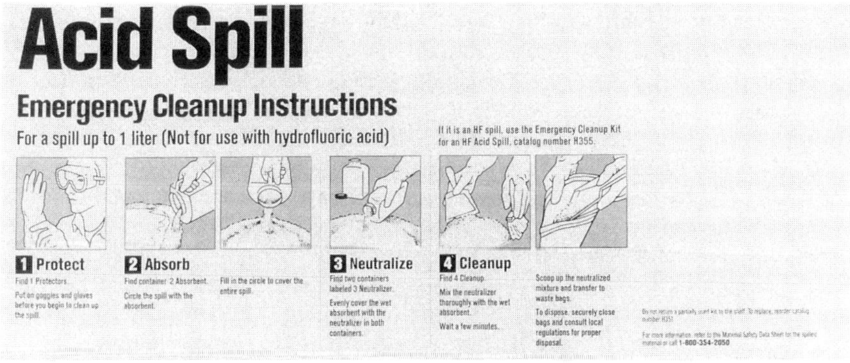

Figure 2. Simplified instructions on the inside lid of the package.

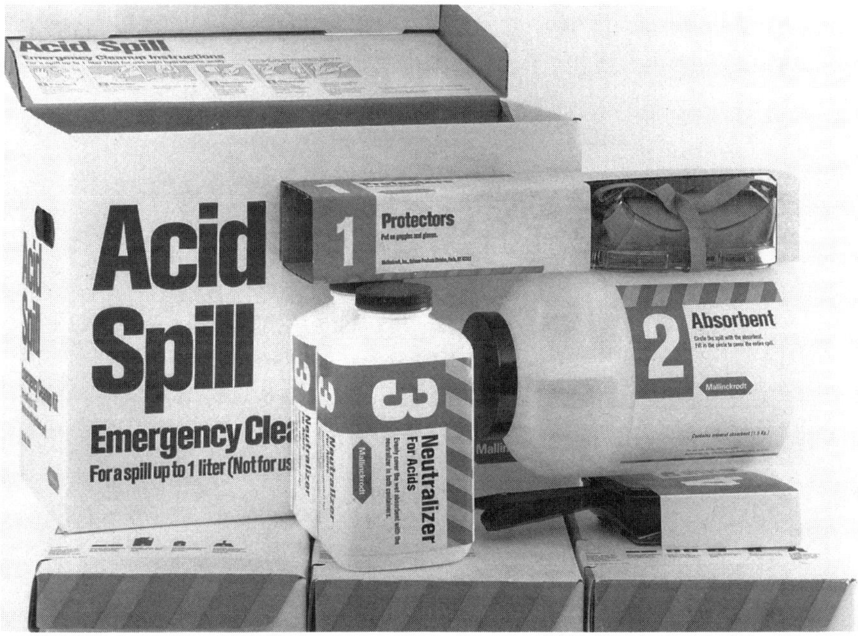

Figure 3. Each component is numbered and contains instructions for use.

weather cords and trouble lights, Carol saw the need for a packaging system that would help them remain predominant in the extension cord industry.

Because extension cords are a commodity product, an item customers find so simple and familiar that price is the primary feature that interests them, there's little opportunity for simplifying the way people use the products. The focus, then, needed to be on simplifying the way people choose the products.

Although consumers feel that extension cords are familiar products, there are some relatively complex safety issues involved in buying extension cords. The wrong cord could cause fire or shock. Carol needed a way of getting customers to look beyond price and consider this important safety information when making a purchase. If Carol could communicate this information and help consumers make the correct purchase, they would reduce returns, improve safety, and encourage repeat business.

As we researched how people buy extension cords, we found several factors about the buying situation that suggested the need for simplified information. We found that customers tend to look for commodity products only when they need them. We also found that customers buy commodity products only when they're noticeable. Keeping all this in mind, we needed to develop packaging information that shouted Carol's existence, required relatively little customer interaction and persuaded customers to make the correct purchase.

Figure 4. Simplified packages highlight primary information for consumers.

Helping the customer to make the correct purchase

From our research, we isolated various bits of information typical customers were interested in and ranked them, some as primary, some as secondary. Our idea was to highlight primary information in different ways—thus catching the eye and addressing the need of any customer who happened to walk by.

The new packaging system helps customers find the right cord a number of different ways: by name (which is the dominant typographic message), by application (which is differentiated through color-coding) or by the type of end it has (which is identified as 2-pronged or 3-pronged in graphics set apart from the text). The result of this structuring was an information hierarchy—a system of chutes—that supports differences in individual need and encourages the correct purchase.

The packaging system also expresses technical information in terms that are more familiar and informative to consumers. Take, for example, the gauge rating. All extension cords have one. Typical ratings are 12, 14, and 16. A 12-gauge cord is more powerful and more expensive than a 14-gauge cord—probably the reverse of the customer's expectation. To eliminate this confusion, we replaced numbers with names. Now, as the cost and power go up, the names go up in value too. The new names speak in everyday terms and emphasize capacity: medium, high, and super versus 16-, 14- and 12-gauge.

We also made sure that Carol's name, the product name, and other primary bits of information were readily apparent. But we wanted to maintain a flexible structure so that customers can easily climb into the copy or randomly read the package without becoming confused. We had to maintain a flexible information system and continue to provide useful information while still *selling* the product. Our dilemma—how to promote without cluttering the language? We discovered an elegant solution. Our research showed that customers found benefits ("this extension cord can be used outdoors") considerably more appealing than features ("this extension cord is triple insulated"). Therefore, the packaging became benefit-oriented, putting information in terms that are more meaningful to consumers. The result for Carol Cable was packaging that explained some complicated ideas simply, helping consumers choose the right product.

CONCLUSION

Although we're pleased that these projects have been successful for our clients, financially and otherwise, that's not the reason for writing about them. Their real success is not what they do, but what they suggest:

The applications of simplified communication are indeed far-reaching. These two examples alone suggest quite a range—from supporting *using* to supporting *choosing,* from industrial to consumer products, from technical to general audiences. But that only scratches the surface. It's only as we push new ways of applying simplified communication and new document forms that we'll discover the real possibilities.

Simplified communication should be considered a process, not a result. The process of simplification should move beyond the surface level of simplifying language. We're beyond the point where it's acceptable to try to simplify a user manual completely separate from the product itself. Simplified communication should be applied at the appropriate interface point—whether it involves simplifying the process by which someone uses a product, or chooses a product, or both.

Writing Backwards: The Use of Visual Models in Writing

G. Reed Agnew
Agnew Moyer Smith Inc., Pittsburgh

The use of visual models in writing technical material can affect the order of the entire writing process — in effect, reversing it. A single case study is used to explain how a model is developed, what purposes it serves, and how it affects the writing process. A few tips on using visual models in writing are given.

Most people would agree that it's good to amplify a piece of writing with visual devices such as diagrams or illustrations. Usually, the sequence of performing these tasks is to write first, then illustrate. Often this process results in visuals that merely embellish the text. Reversing the sequence of writing and illustrating can significantly improve a writer's ability to communicate concepts effectively, but will also change the writing process. Because using this technique essentially reverses the usual writing process, I call it Writing Backwards. A project completed by our office (Agnew Moyer Smith, Inc.) clearly illustrates the technique of writing backwards. This project dealt with the need to explain how an integrated information system would work and how it could be built.

Planning an office of the future

An integrated information system is sometimes referred to as "the office of the future." It is a system that links together the latest office technologies — word processors, computers, telephone services, video, audio visual, laser printers, intelligent copies, and typesetters. Once linked, the synergism of the total system profoundly changes the way office work is accomplished. Our client, Westinghouse corporate headquarters, was planning to build such a system.

To begin, the company had formed a task force composed of experts in each area of technology. Each member of the group researched available products and techniques within his or her own area and then completed a detailed report explaining a recommended approach. Together, in many meetings, the task force worked out many of the problems of connecting these often disparate technologies. The result of this initial effort was two thick binders full of technical papers, each paper explaining a specific area of technology. The binders were filled with complicated system diagrams and the text was riddled with unexplained technical jargon like asychronous, gateways, and COM.

Reprinted by permission of the publisher, *Visible Language,* Rhode Island School of Design, Providence, RI 02903.

The problem

The problem was that, for the most part, these reports were only understandable by a technical person — and this wasn't the audience. The audience was the company's top-level managers who were going to be asked to approve the expenditures for the system, and middle managers whose departments were going to be the system users. For these audiences, it was important to show: How the system would work, what benefits it would offer, how much it would cost, and how it could be built. These questions had to be answered simply but without compromising technical integrity, and in a way which could be easily and quickly comprehended.

Our office was asked to solve this communications problem. The office was founded by graphic designers — visual thinkers, skilled in design, layout, typography, and the visual organization of publications. Yet the office is committed to the idea that designers cannot merely lay out and amplify writers' ideas and text — that visual and verbal thinking cannot be separated. Our goal is to integrate writing and design and to recruit other designers who are interested in dealing with communication problems — designers who are also writers. The make-up and attitudes of our office strongly influenced our approach to explaining the integrated information system. Here's what we did:

A case study

First, we read and reread the technical reports, writing notes, making queries, and checking other references while we tried to understand the system and its components. Next, we talked to selected task force members about their particular technical areas. We clarified misunderstandings and learned about their primary concerns and objectives. We also asked them about their perceptions of the audience's information needs and what areas seemed particularly confusing or unclear. Then, we went away and thought.

Developing a visual model

After we felt comfortable with the subject and information needs, our first productive task was to develop a visual model of the integrated system we thought they were trying to build. A visual model is a diagram, chart, illustration, or matrix that shows many parts of the subject simultaneously. Once constructed, a good model enables a viewer to comprehend a complete concept or idea all at once.

In this case, the model was an axonometric diagram which represented all the system's components in the context of three typical floors of an office building. We used this type of diagram because both task force members and their audience seemed confused about where equipment would be placed and how much impact the system would have on the physical arrangement of offices. We have also learned that readers naturally respond to and understand axonometric drawings with representations of real equipment more readily than they do to two-dimensional flow diagrams filled with arrows and abstract shapes like boxes and circles.

Our first presentation of this model to the task force was enlightening for everyone. "Yes," they said, "that is pretty close to representing the system we're trying to build." Then the various technical experts began to discuss the details of the model (Figure 1). To everyone's surprise, it became apparent that there were some holes in the system, some incompatibilities, unforeseen links, and unsolved problems. In the next few weeks, there were many meetings and in each one this visual model became the focal point of discussions. Copies of the model were used to collect and consolidate new information, to discuss changes, and to think through the construction sequence. It was the first time members of the task force had a device to communicate among themselves, to relate their separate technologies, and to consolidate their ideas. The result of these sessions was a revised diagram that everyone agreed was an accurate representation of the system. In addition, we color-coded a series of these diagrams to show how components would be added over a five-year period.

Once this task was completed, we were well on our way to developing a simplified report for the intended audience. Now we could explain the system, benefits, costs, and construction sequence in terms of the visual model. In effect our writing task now was to describe how the model worked — not to rewrite and edit the binders full of complex technical papers. Our writing task would be easier and we had a strong organizing principle to follow.

Figure 1. The visual model became the focal point for conducting task force discussions, for collecting new information, and resolving problems in the system design and construction plans.

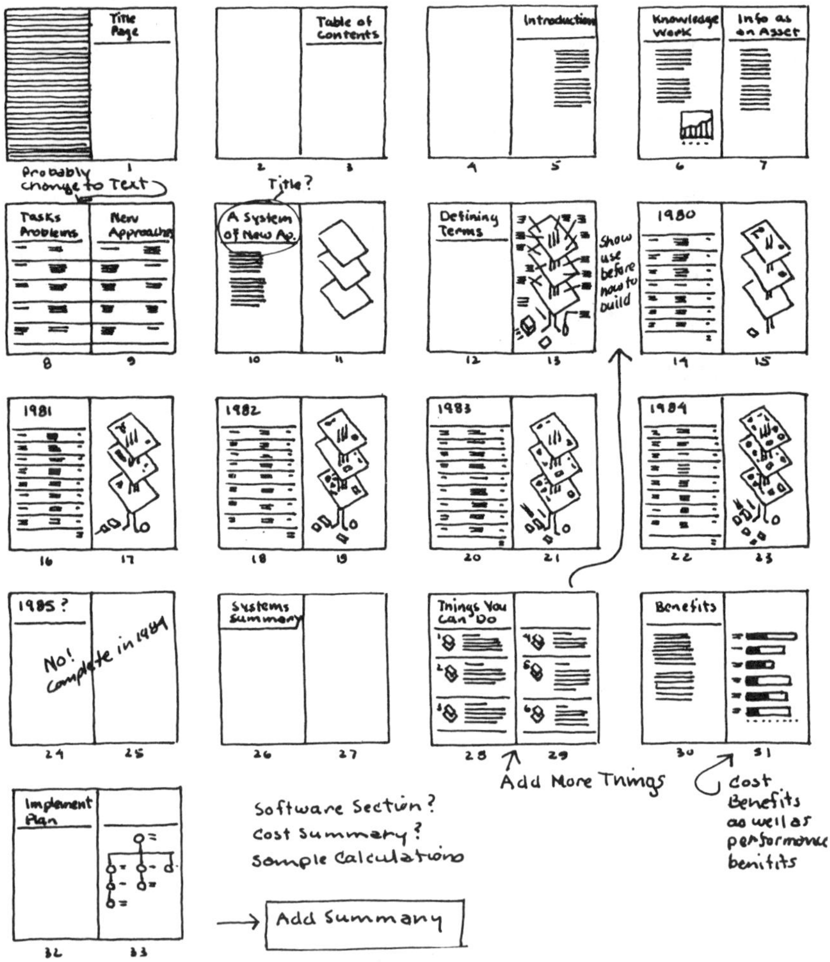

Figure 2. The visual outline was developed to plan the organization and content of the book. It became the guide for detailed writing.

Creating a visual outline

To begin writing, we created another kind of visual model — a model of the publication we envisioned. This model was much less refined than our system diagram, but was absolutely essential in planning how we would explain the system to our readers. We call this type of model a visual outline (Figure 2). It's simply a small sketch of two-page spreads with indications of what will be explained on each page or spread. It's a combination of an outline and a page

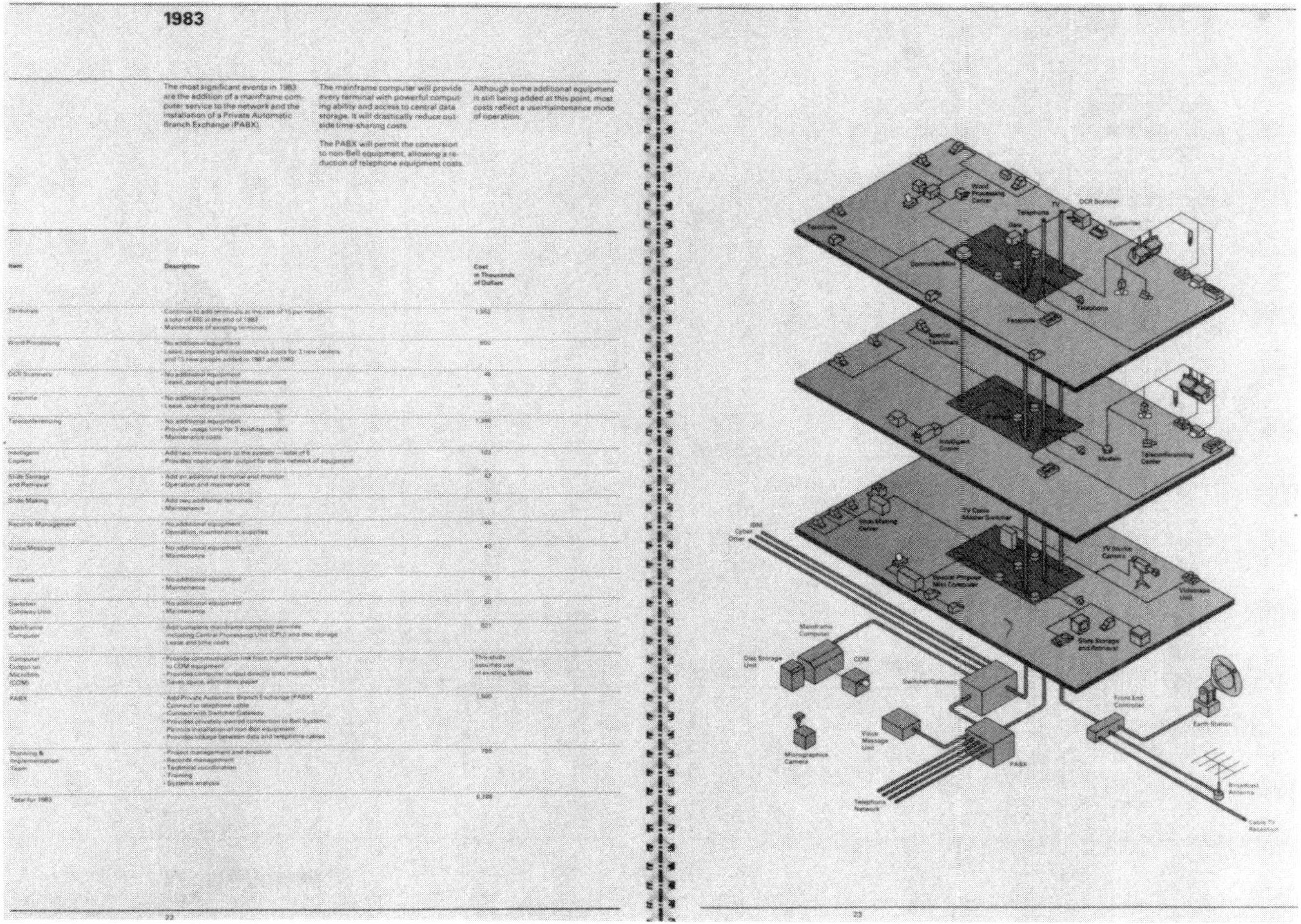

Figure 3. A series of pages similar to this two-page spread use the diagram to show how the system would be built over a 5-year period. The left-hand page of each spread contains a table explaining what new components were added and how much they cost.

Figure 4. Use scenarios were developed which describe step by step how 15 office tasks such as filing, document sharing, video conferencing, or electronic mail would be performed using the new system. Small diagrams use color to indicate which system components are used in each scenario.

1983

The most significant events in 1983 are the addition of a mainframe computer service to the network and the installation of a Private Automatic Branch Exchange (PABX).

The mainframe computer will provide every terminal with powerful computing ability and access to central data storage. It will drastically reduce outside time-sharing costs.

The PABX will permit the conversion to non-Bell equipment, allowing a reduction of telephone equipment costs.

Although some additional equipment is still being added at this point, most costs reflect a use/maintenance mode of operation.

Item	Description	Cost in Thousands of Dollars
Terminals	· Continue to add terminals at the rate of 15 per month — a total of 600 at the end of 1983 · Maintenance of existing terminals	1,552
Word Processing	· No additional equipment · Lease, operating and maintenance costs for 3 new centers and 15 new people added in 1981 and 1982	600
OCR Scanners	· No additional equipment · Lease, operating and maintenance costs	46
Facsimile	· No additional equipment · Lease, operating and maintenance costs	25
Teleconferencing	· No additional equipment · Provide usage time for 3 existing centers · Maintenance costs	1,346
Intelligent Copiers	· Add two more copiers to the system — total of 5 · Provides copier/printer output for entire network of equipment	103
Slide Storage and Retrieval	· Add an additional terminal and monitor · Operation and maintenance	42
Slide Making	· Add two additional terminals · Maintenance	17
Records Management	· No additional equipment · Operation, maintenance, supplies	45
Voice/Message	· No additional equipment · Maintenance	40
Network	· No additional equipment · Maintenance	20
Switcher/Gateway Unit	· No additional equipment · Maintenance	50
Mainframe Computer	· Add complete mainframe computer services including Central Processing Unit (CPU) and disc storage · Lease and time costs	621
Computer Output on Microfilm (COM)	· Provide communication link from mainframe computer to COM equipment · Provides computer output directly onto microfilm · Saves space, eliminates paper	This study assumes use of existing facilities
PABX	· Add Private Automatic Branch Exchange (PABX) · Connect to telephone cable · Connect with Switcher/Gateway · Provides privately owned connection to Bell System · Permits installation of non-Bell equipment · Provides linkage between data and telephone cables	1,320
Planning & Implementation Team	· Project management and direction · Records management · Technical coordination · Training · Systems analysis	766
Total for 1983		6,788

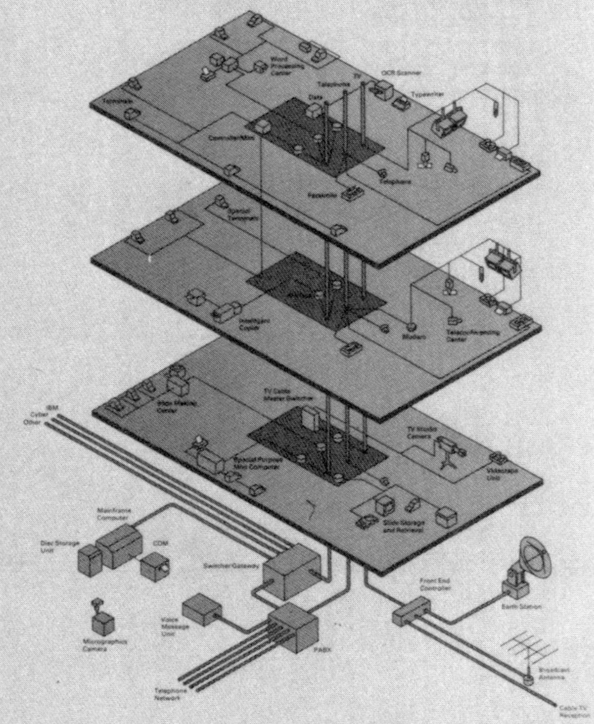

23

plan. We have found that outlining in this way helps to ensure proper page breaks, aids in planning the inclusion of visual material, allows us to assess space problems early, and ensures proper pacing of the presentation. We've found visual outlines to be much more useful than the traditional hierarchical outlining technique we all learned in school. In addition, visual outlines help the clients "see" their publication, and approve the concept and organization before detailed writing begins. In this project, the visual outline helped us plan how we would integrate text with our axonometric diagram. It showed where writing was needed to fully explain an idea and where the diagram could carry the full burden of communication.

Writing the text

At this point, we began writing the text. And it was easy because, for the most part, the text was interpreting and amplifying the visual model. It's much easier to write about a subject when you have a coherent, visible conception of it in front of you. The publication explained the complete system in only 32 pages. The amount of text was limited; information was often organized into matrices or tables instead of text to make it more accessible and inviting (Figures 3 and 4). The system diagram was easy to understand as well as visually engaging. Today, our client is using an integrated information system like the one described in this book.

Writing the usual way

The typical approach to most writing assignments usually follows this pattern: (1) *Interview and set criteria.* The writer interviews the client and establishes basic criteria about the purpose, audience, schedule, cost, and general approach to the assignment. (2) *Research and collect information.* A reference file will begin to develop. (3) *Organize and categorize.* Already some basic decisions which will affect writing are beginning to be made as information is sorted and eliminated. (4) *Outline.* The writer orders the information into a basic hierarchy using words and phrases to describe the content of each category. (5) *Write.* Now the writer "fleshes out" the outline and writes text supporting the order that has been established. (6) *Illustrate and visualize.* The writer reads through to see what points could be illustrated or where tables, graphs, or charts could be included to offer the reader some visual relief. (7) *Lay out.* Finally, the writer decides how the text and illustrations can be organized and arranged in pages.

Writing backwards

Agnew Moyer Smith, Inc. does not follow the traditional approach to writing. The first three steps are the same, but we reverse the last four steps. We begin with visualization, then we simultaneously complete layout and outlining (our visual outline), and finally we write. Our process is backwards.

We have found this to be an extremely useful technique — particularly with technical subjects or with subjects that are organizationally complex. Again and again, visual models have proven to be the bridge between our clients' inability to communicate knowledge clearly and our own ability to understand the ideas and concepts they want to convey. Discussion, change, and modification of models seems to be much easier and less threatening than rewriting or editing textual explanations. And once agreed to, models and visual outlines can readily become a strong organizing principle which the text can easily support.

Our office has used this process on a number of projects — each one requiring a different type of model. One project used a timeline model to unravel, organize, and explain the complex task of relocating families transferred from one company location to another. A well-structured, engaging, and easily read book organized around the timeline made a complex and potentially disruptive task easy. Other projects have used visual models to explain how animals behave, how inventory financing works, what components a manufacturer supplies to the mass transit market, how a project is structured within a matrix organization, and how an automated information retrieval system works. In all these examples, the model was created first. And in all cases, the existence of a model helped our clients clarify fuzzy ideas, eliminated the need for a lot of words, and provided a strong organizing principle for writing.

Integrating text with visual models

Although well-conceived visual models can assume a good share of the communications burden, it is still important to work hard on the clarity of the text, captions, and call-outs that accompany the model. We have found that the existence of a strong model affects writing style. Projects with strong visual models have less main text, and rely more on the supporting layers of text such as call-outs and captions. In addition, these supporting levels of text tend to be written in a terse, telegraphic style. Finally, incorporating visual models changes the traditional structure of pages. No longer will a page contain long columns of text which makes occasional references to remote supporting figures. Text and visuals must be fully integrated.

Although developing visual models early in the writing process is a useful and challenging approach to writing, it's not for everyone. Some people cannot think visually; they cannot represent a concept with even the crudest sketch. Still, many writers are perfectly capable of developing conceptual visual models. One does not have to have a background in illustration or graphic design to use this approach.

Tips on using visual models

To conclude, I can offer a few tips on using visual models. The following techniques have proven useful in most of the projects we have completed.

Don't discard or put aside explanatory sketches. Rough diagrams drawn on the back of an envelope or a napkin during a discussion may contain the beginnings of a good idea for a visual model.

Look for metaphors and comparisons as clients explain ideas or concepts. These casual comparisons can lead to a strikingly clear visual concept.

Don't worry about the artistic quality of your first sketches. If you have a good idea that truly helps explain a concept, you can find a designer or illustrator to work out the details later.

Remember your audience. Be sure your model is not too complex or technical for a lay audience, but don't oversimplify or talk down to a more knowledgeable reader.

Create models early in the process. That's when they do the most good for you and your client. They can be a valuable tool in learning a subject and resolving misunderstandings. Don't use models to simply embellish your text after it is completed.

Don't expect the model to replace clearly written, well-organized text. Models do not excuse poor writing.

The next time you're trying to explain something, don't plunge right into outlines, text and words. Pause for a while and think about how you might approach a visual explanation. It could be a rewarding experience.

Contributors

G. Reed Agnew is one of the founding principals of Agnew Moyer Smith, a graphic design and communications consulting office. Prior to forming this office in 1980, he spent ten years with the Westinghouse Corporate Design Center. He is specifically interested in the integration of language and organization techniques with visual concepts. His work has won design awards from the American Institute of Graphic Arts and the Society of Typographic Arts and has been published in *Print Magazine* and in the books *Graphic Design USA 1* and *Graphic Design USA 2*.

Hiraku Amemiya is a document design manager for High Technology Communications in Tokyo and is managing editor of *The Eastern Communicator (TEC),* a semi-annual newsletter published by HTC. A founding member of several fledgling professional societies, he is actively engaged in efforts to win recognition among Japanese manufacturers and users of the role and importance of technical writing. His major publications include *The Complete Guide to Writing Software User Manuals* (a Japanese translation of the book by Robert McGehee), *My First Apple,* and *JCALC Users' Guide.*

Ronald G. Amodeo was one of the founding associates of the Information Design Resource Group at Fitch RichardsonSmith. He is now manager of Technical Information at Printronix. He received his M.A. in professional writing from Carnegie Mellon University.

Sandra J. Bond is a senior writer and editor at Carnegie Mellon University's Software Engineering Institute, a federally funded research and development center established to provide leadership in transitioning software from state-of-the-art to state-of-the-practice. She has an M.A. in professional writing from Carnegie Mellon. Her particular interest is in writing computer documentation in plain English.

Betsy A. Bowen is an assistant professor of rhetoric and composition at Fairfield University. She received her Ph.D. in rhetoric from Carnegie Mellon University, after having studied at the University of New Hampshire and the Bread Loaf School of English. Her research examines the cognitive processes of collaborative revision and peer tutoring.

Elizabeth H. Curtin earned a D.A. in rhetoric from Carnegie Mellon University. She teaches English at Chartiers Valley High School in Bridgeville, Pennsylvania.

THOMAS M. DUFFY is an associate professor in instructional systems technology and a research associate in learning resources at Indiana University. He has contributed articles on plain language to books and journals, and he is coeditor of the book *Designing Usable Texts.* His award-winning article, "Expertise and the Process of Developing Technical Manuals," appeared in *Technical Communication.* More recently Duffy has been working on the design of information for online presentation and has written "Hypermedia and Education: Where is the Match," a chapter in a forthcoming book. He is currently writing a book, *Online Help: Design and Evaluation.*

GAIL DYKSTRA is the director of the Canadian Law Information Council's Plain Language Centre, where she is responsible for all of the Council's public legal education activities. She has written several papers on plain language, including "Language and the Law," which served as the impetus for the creation of the Plain Language Centre, and she is the author of articles on all aspects of public access to information on the law and public legal education. Dykstra has also served as editor of *CLIC's Abstracts on Public Legal Education, Legal Materials Letter, Network News,* and a number of newsletters for associations with an interest in legal information.

ROBERT D. EAGLESON is associate professor of English at the University of Sydney in Australia. He helped draft the first plain English insurance policy in Australia in 1976 and subsequently has served as a consultant to many private and government organizations. As Special Adviser on Plain English to the Australian government, Eagleson prepared a manual on plain English that will form the basis of a nationwide government training program in official writing. In 1986 and 1987 he served on the Law Reform Commission of Victoria as commissioner in charge of the plain English reference, which was concerned particularly with the drafting of legislation and which led to the report *Plain English and the Law* (Melbourne, 1987). He is author and coauthor of a number of books and articles on modern English language and Shakespeare's language.

IRENE A. ETZKORN is a senior vice president at Siegel & Gale. As director of the company's Simplified Communications Group, she coordinates the group's activities and participates in major simplification projects. Her projects include developing simplified forms, manuals, instructions, and account statements for a broad range of clients, including Citibank, Xerox, GTE, MCI, and Shearson Lehman Hutton. She has also worked for a number of government clients, including the New York City Department of Finance and Revenu Québec. In 1983, she served as editor-in-chief for the Army Task Force of the President's Private Sector Survey on Cost Control. She holds an M.A. in professional writing from Carnegie Mellon University.

MICHAEL J. FLOREAK is director of Information Design at Fitch RichardsonSmith in Boston. He works closely with graphic and product designers to simplify communications and products for clients such as Philips,

Xerox, and Black & Decker. He received his M.A. in professional writing from Carnegie Mellon University.

Carl Felsenfeld is a professor of law at the Fordham University Law School. He specializes in the regulation of banks and other financial institutions. He was a charter member of the Federal Reserve Consumer Advisory Council and serves as a delegate to the United Nations Committee on International Trade Law. He was primarily responsible for Citibank's plain English forms breakthrough in 1975.

Lawrence T. Frase is Distinguished Member of Technical Staff at AT&T Bell Laboratories. He has published research on human reasoning, reading, writing, and document design. While on leave from Bell Laboratories, he was head of the learning division of the National Institute of Education in Washington, D.C. Currently he is working on applications of computers to human information processing. He is a fellow of the American Psychological Association and holds a Ph.D. in educational psychology from the University of Illinois.

Debra Graham is a technical writer for Software Architecture and Engineering. Her major responsibility is documentation for her firm, a private software company. She received an M.A. in professional writing from Carnegie Mellon University.

John R. Hayes is professor of psychology and co-director of the Communication Design Center at Carnegie Mellon University. He does research in written composition, problem solving, and creativity, and he is the author of *Cognitive Psychology: Thinking and Creating* and *The Complete Problem Solver*. Hayes is a fellow of the American Psychological Association and the American Association for the Advancement of Science.

Thomas N. Huckin is associate professor of English and the University Writing Program at the University of Utah. He received his Ph.D. in linguistics from the University of Washington and has taught technical writing and English-as-a-second-language at the University of Michigan and at Carnegie Mellon University. He is coauthor of *Principles of Communication for Science and Technology.*

Carol J. Janik is a technical writing consultant specializing in software documentation. She is the author and coauthor of seven computer manuals and has won two awards from the Pittsburgh chapter of the Society for the Technical Communication. Janik earned her M.A. in professional writing from Carnegie Mellon University.

Barry Jereb is supervisor of Industrial Computer Hardware and Software Documentation at the Allen-Bradley Company, Industrial Computer and Communications Group. He has contributed articles to *Business Marketing,*

Visible Language, and *Rhetoric Society Quarterly* and has been a panel member for both international and regional conferences of the Society for Technical Communications. In addition, his work has won international and regional awards from the Society for Technical Communications. He received his M.A. in professional writing from Carnegie Mellon University.

Nina H. MacDonald has been with AT&T Bell Laboratories since 1976, where she has studied the role of pauses in sentence comprehension and worked on the development and dissemination of information about *UNIX*™ and WRITER'S WORKBENCH™ software. She is currently working on the design and implementation of user interfaces and front-ends for her organization's Libraries and Information Systems Center and programming a new current-awareness bulletin production system. She holds a Ph.D. in Linguistics from the University of Michigan.

Lance A. Miller is technical assistant to the Vice President for Technology at IBM Federal Systems Division. Prior to that, he was at IBM's Watson Research Center, where he conducted and managed research in human-computer interaction, first in a long-term study of the cognitive problems of programming and then in the development of an expert text-critiquing system called EPISTLE. His particular area of concern is the development of natural language processing technologies to support improved natural language interfaces to computers as well as text generation application. He holds a Ph.D. in experimental psychology from the University of Wisconsin (Madison).

Dr. Janice C. (Ginny) Redish is a vice president of the American Institutes for Research and is director of AIR's Document Design Center. Redish is a linguist who has published extensively on the characteristics of technical and legal language, on readability, and on organizing documents to make information accessible. She has trained hundreds of writers to change their approach to writing from a traditional, nominal style to a reader-oriented, verb-based style.

Susan Rosen is a technical writer at the Document Design Center and a doctoral candidate in English at the University of Maryland.

Karen A. Schriver is an assistant professor of rhetoric and document design at Carnegie Mellon University. She is co-director of the Master of Arts in professional writing program and teaches undergraduate and graduate courses in integrating visual and verbal text and planning and testing documents. Her research focuses on the cognitive and social processes involved in planning and revising academic and non-academic texts. She is also investigating conceptions of theory-building in rhetorical, literary, and composition studies. She is a research associate and publications coordinator to the national Center for the Study of Writing and is a co-recipient of the 1987 Richard Braddock Award. She holds a Ph.D. in rhetoric from Carnegie Mellon University.

ERWIN R. STEINBERG is Thomas S. Baker Professor of English and interdisciplinary studies at Carnegie Mellon University and codirector of that university's Communication Design Center. He organized the seminar at which many of the papers appearing in this volume were presented. He is coauthor of *Communications in Business and Industry* and coeditor of *Cognitive Processes in Writing*. He has also written and edited books on literature and has contributed articles to journals, including *Journal of Modern Literature, Modern Fiction Studies, Advanced Management, American Banker, Journal of Advanced Composition,* and *Written Communication*. Steinberg is a former chairman of the Conference on College Composition and Communication and has served for many years as a communications consultant in industry.

JOYCE HANNAH SWANEY is an independent documentation consultant, specializing in software user manuals. Prior to starting her own business, she was a technical writer and research assistant at Carnegie Mellon University's Communications Design Center and supervisor of documentation at PERQ Systems in Pittsburgh, PA. She received a B.A. in writing from the University of Pittsburgh.

LORRAINE K. WILKIN is director of Information Design at Fitch RichardsonSmith in Worthington, Ohio. Her interests include designing user interfaces and simplifying complex information systems. She is a member of the Society for Technical Communication and the Human Factors Society. She received her M.A. in professional writing from Carnegie Mellon University.

JOSEPH M. WILLIAMS is a professor in the departments of English and linguistics at the University of Chicago. He has published books and articles on stylistics, the history of the English language, teaching composition, and social linguistics. He wrote *Origins of the English Language,* which has been adopted as the text to accompany the recently produced Public Broadcasting System series on the English language, and *Style: Ten Lessons to Clarity and Grace,* which is widely used in advanced writing programs. Williams has also consulted extensively with corporations, the Federal government, and major law firms.

Index